Super Smart Science Series
COLLECTION
Books 1 through 5

Cellular Biology:
Organelles, Structure, Function

Chemistry:
The Atom and Elements

Neurology:
The Amazing Central Nervous System

Astronomy:
The Solar System

Anatomy & Physiology Part 1:
Bones, Muscles, and The Stuff
That Connects Bones and Muscles

Dedicated to:
MOM AND DAD

Super Smart Science Series Combo Book : 1 through 5
ISBN #: 978-1-941775-14-1
April Chloe Terrazas, BS University of Texas at Austin
© 2015 Crazy Brainz Publishing
Visit us on the web! www.Crazy-Brainz.com

Cover design, illustrations and text by: April Chloe Terrazas

Cellular Biology:

Organelles, Structure, Function

Ages 0-75

Cell Membrane

Vesicle

Lysosome

Mitochondria

Nucleus

Microtubules

DNA

Golgi

Ribosomes

Centrioles

Cytoplasm

Hormone

Membrane Protein

Book ONE of the

Super Smart Science Series

By: APRIL CHLOE TERRAZAS

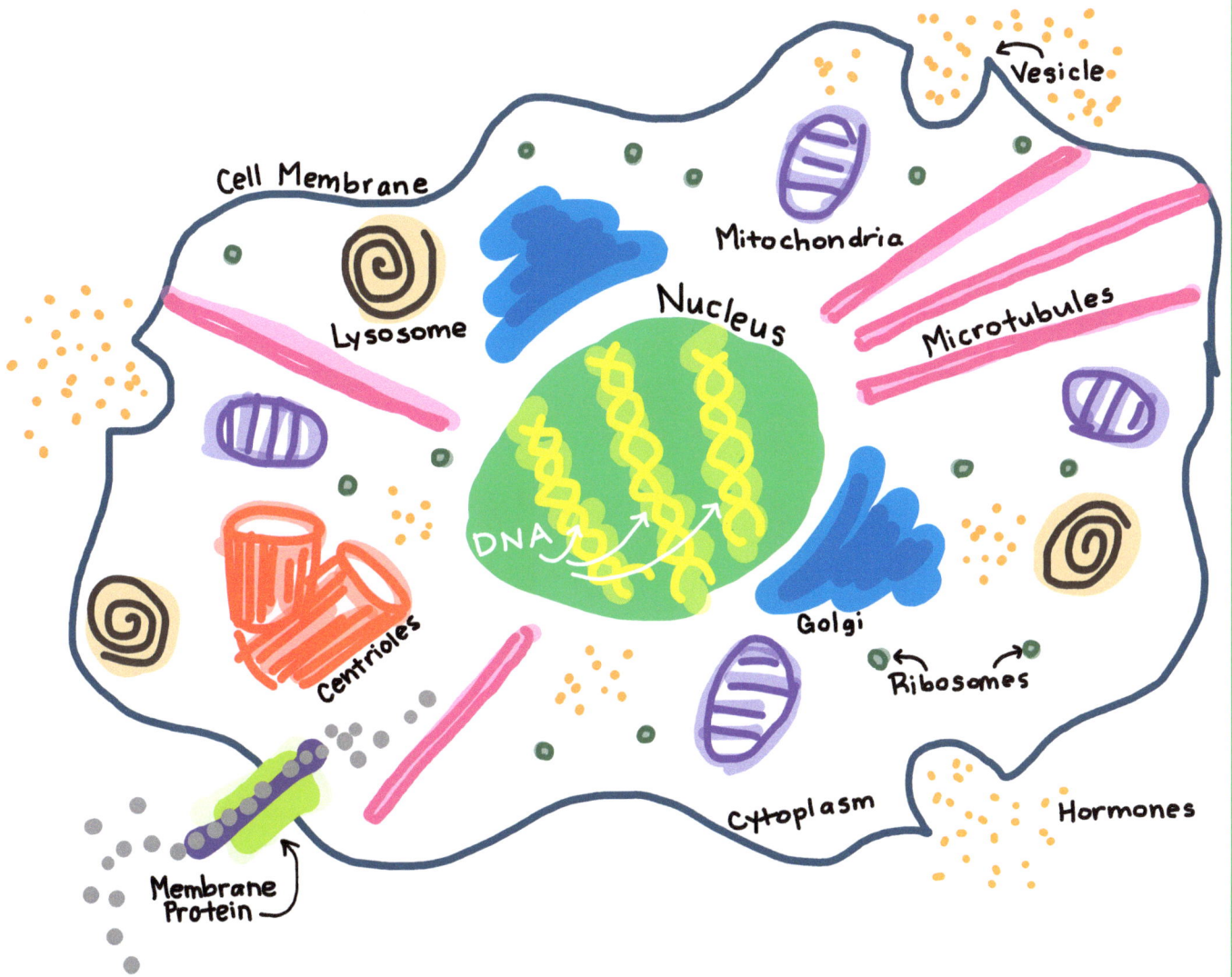

Cellular Biology: Organelles, Structure, Function

Written and Illustrated by: APRIL CHLOE TERRAZAS

Dedicated to my
beloved cousin.
A.J. Ripps
(1987-2012)

Cellular Biology: Organelles, Structure, Function. April Chloe Terrazas, BS University of Texas at Austin.
Copyright © 2012 Crazy Brainz, LLC

Visit us on the web! www.Crazy-Brainz.com

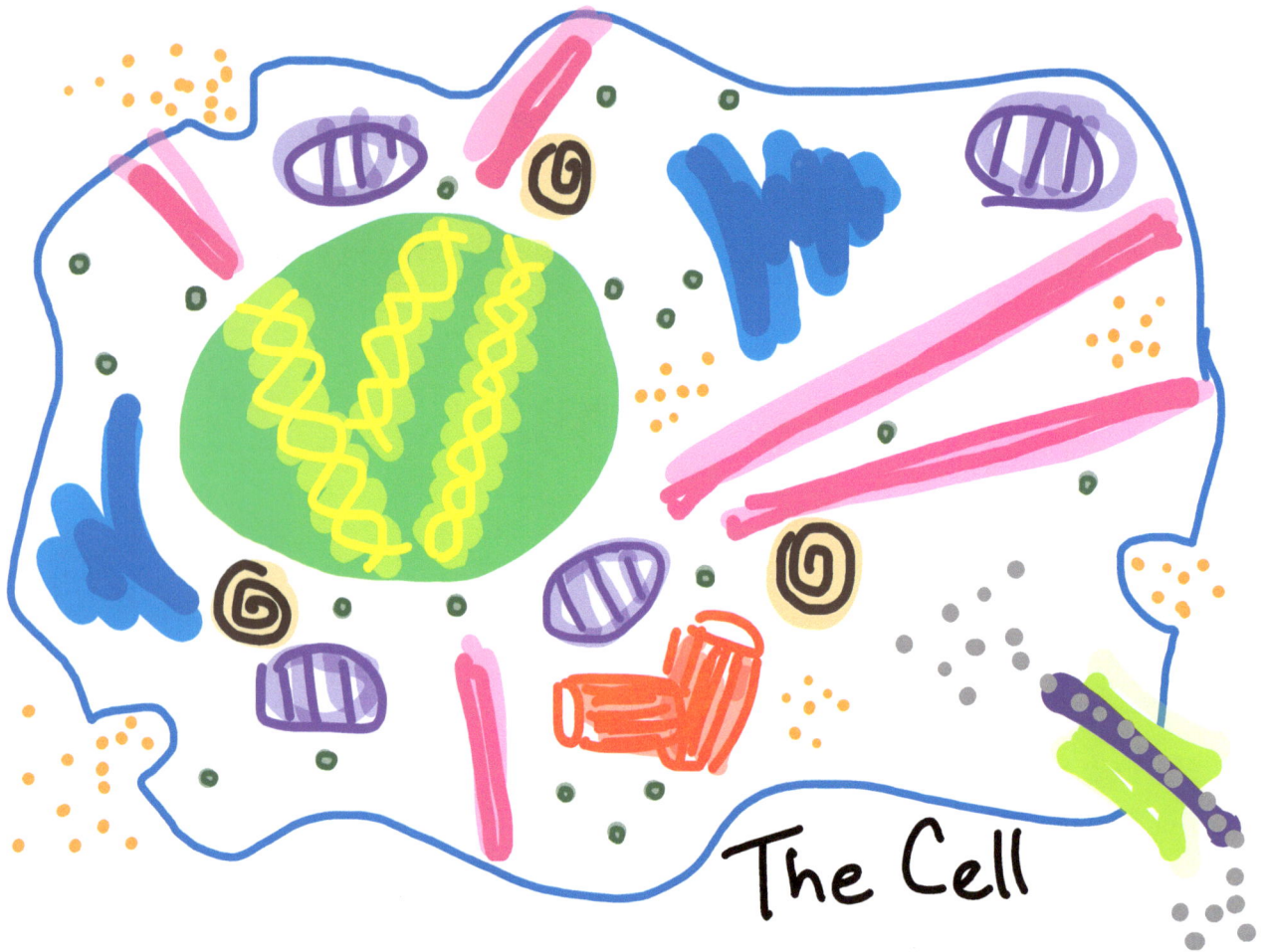

The Cell

This is a cell.

All living things are made of cells.

Cell

There are small parts inside each cell called organelles.

Or-gan-elles

Do you see all of the colored organelles inside the cell?

Turn the page to begin learning about the cell.

You are becoming a cell expert!

Nucleus

Nucleus

DNA

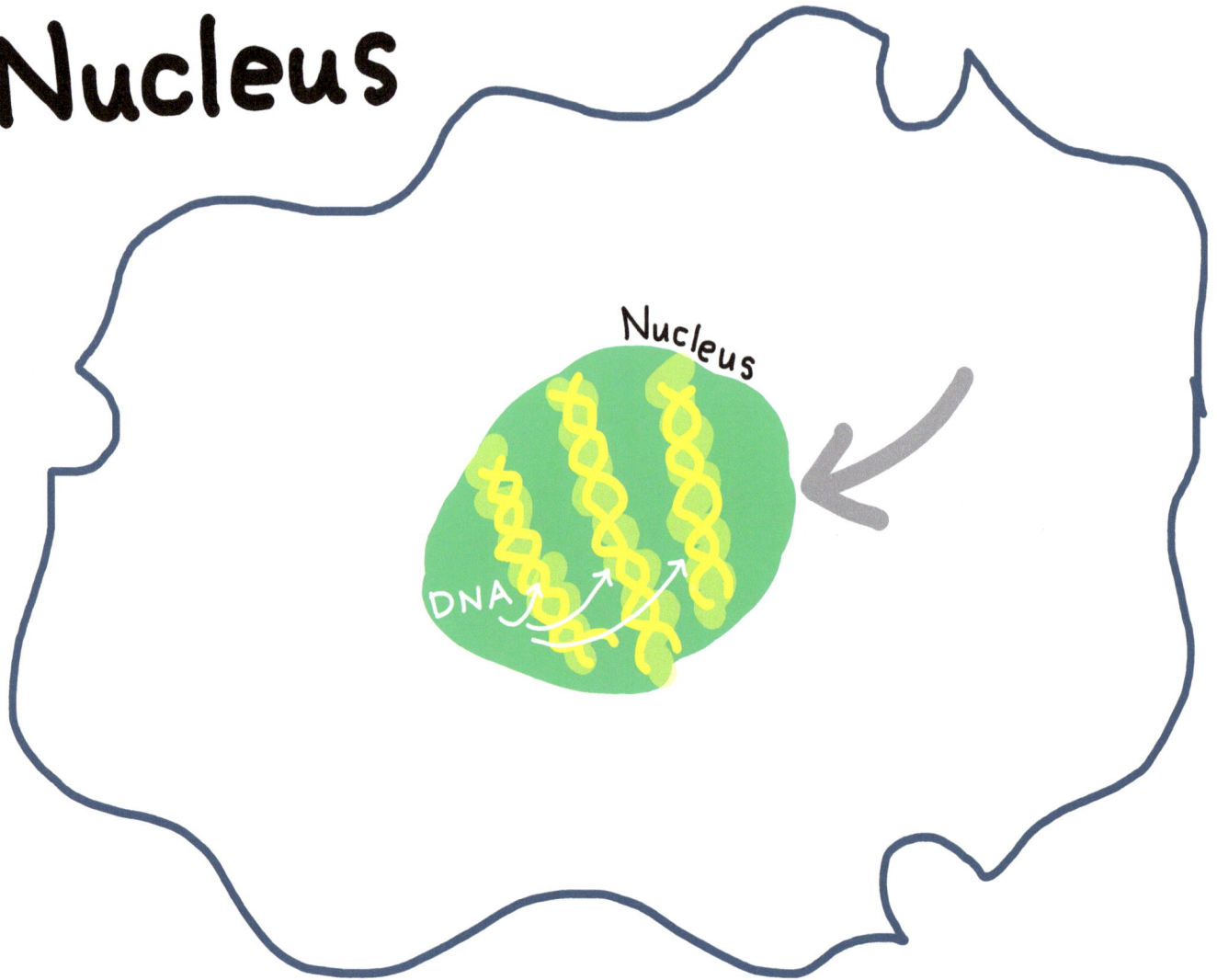

This is the **nucleus**.

Do you see the *yellow* lines in the **nucleus**?

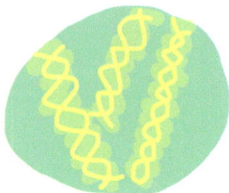

Nu-cle-us

Sound It Out
1. NU
2. KLEE
3. US

DNA is inside the nucleus.

DNA is what <u>makes each person different and special</u>.

The nucleus is like the brain of the cell.

The nucleus <u>controls everything</u>!

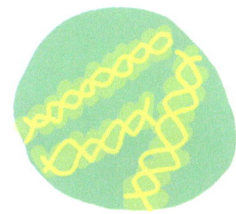

Point to the nucleus and the DNA.

How many strands of DNA do you see?

Centrioles

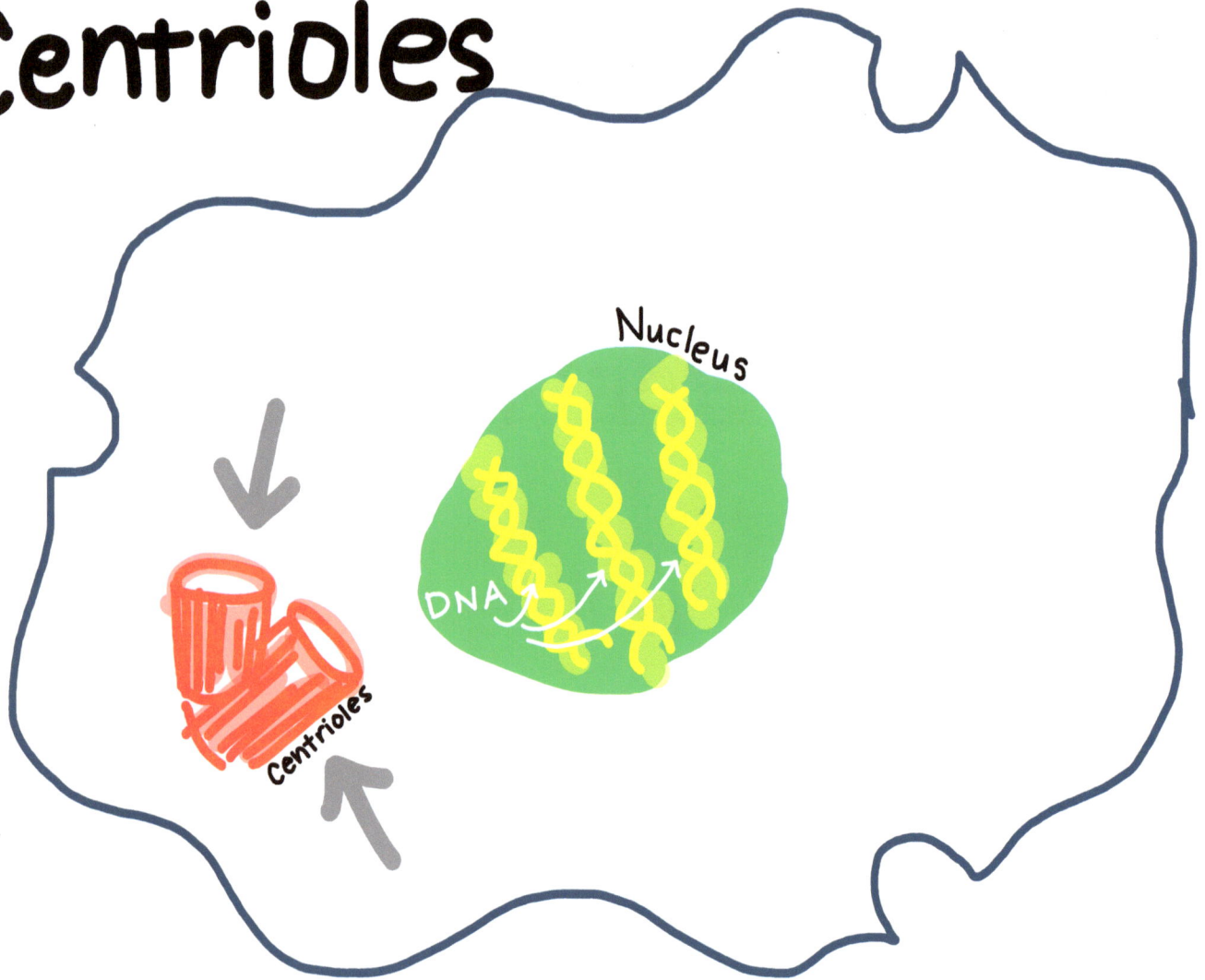

Nucleus

DNA

Centrioles

These are **centrioles**.

Do you see the **centrioles** next to the **nucleus**?

Cen-tri-oles

Sound it Out

1. **SEN**
2. **TREE**
3. **OLS**

Centrioles are organelles inside the cell.

Centrioles help the cell make new cells.

Point to the DNA inside the nucleus.

What does the nucleus do?

Microtubules

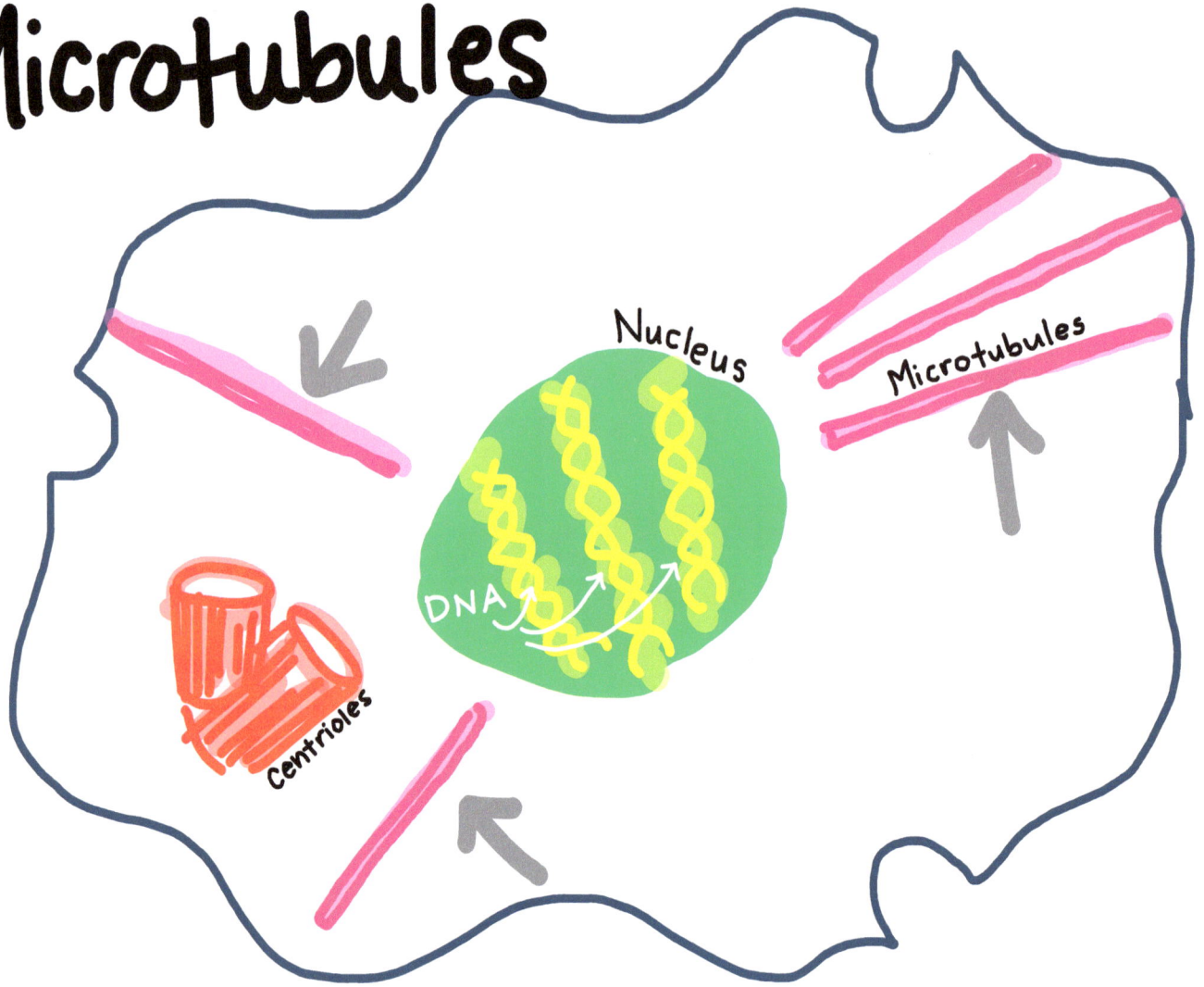

Nucleus

Microtubules

DNA

Centrioles

Mi-cro-tub-ules

Sound It Out

1. MI
2. KRO
3. TOOB
4. ULZ

These are **microtubules**.

Microtubules are organelles inside the cell.

Microtubules <u>move things</u> around inside the cell.

Do you see the *microtubules*?

Can you find the centrioles?

What do the centrioles do?

Where is the *nucleus*?

Mitochondria

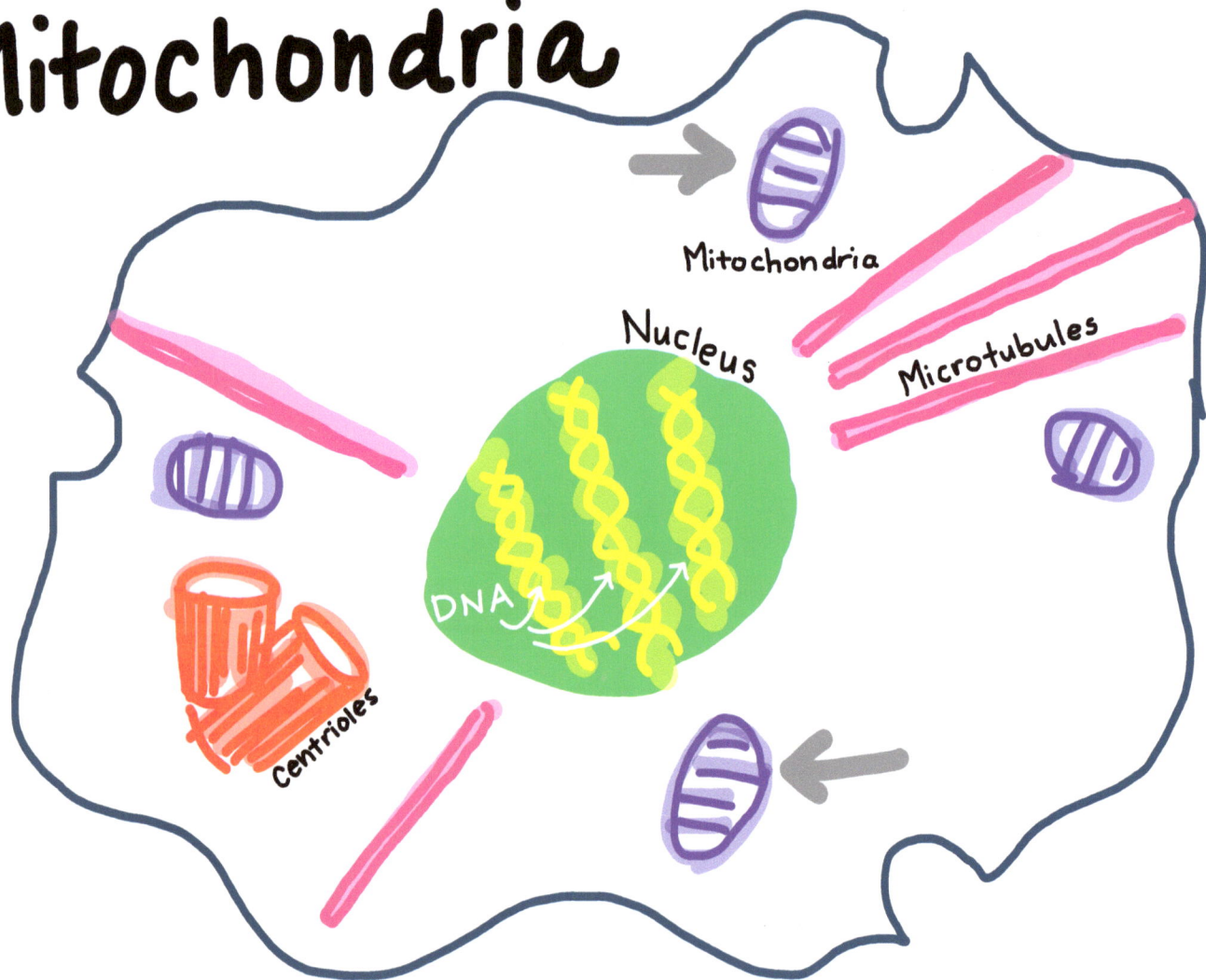

Mitochondria

Nucleus

Microtubules

DNA

Centrioles

Mi-to-chon-dri-a

These are mitochondria.

Mitochondria are **organelles** inside the **cell**.

The **mitochondria** in your **cells** <u>make energy</u> so you can run and play!

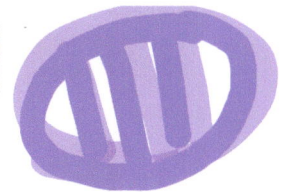

Point to the **mitochondria**.

Where are the **centrioles**? Do you see the **DNA** and *nucleus*?

What do *microtubules* do inside the **cell**?

Lysosomes

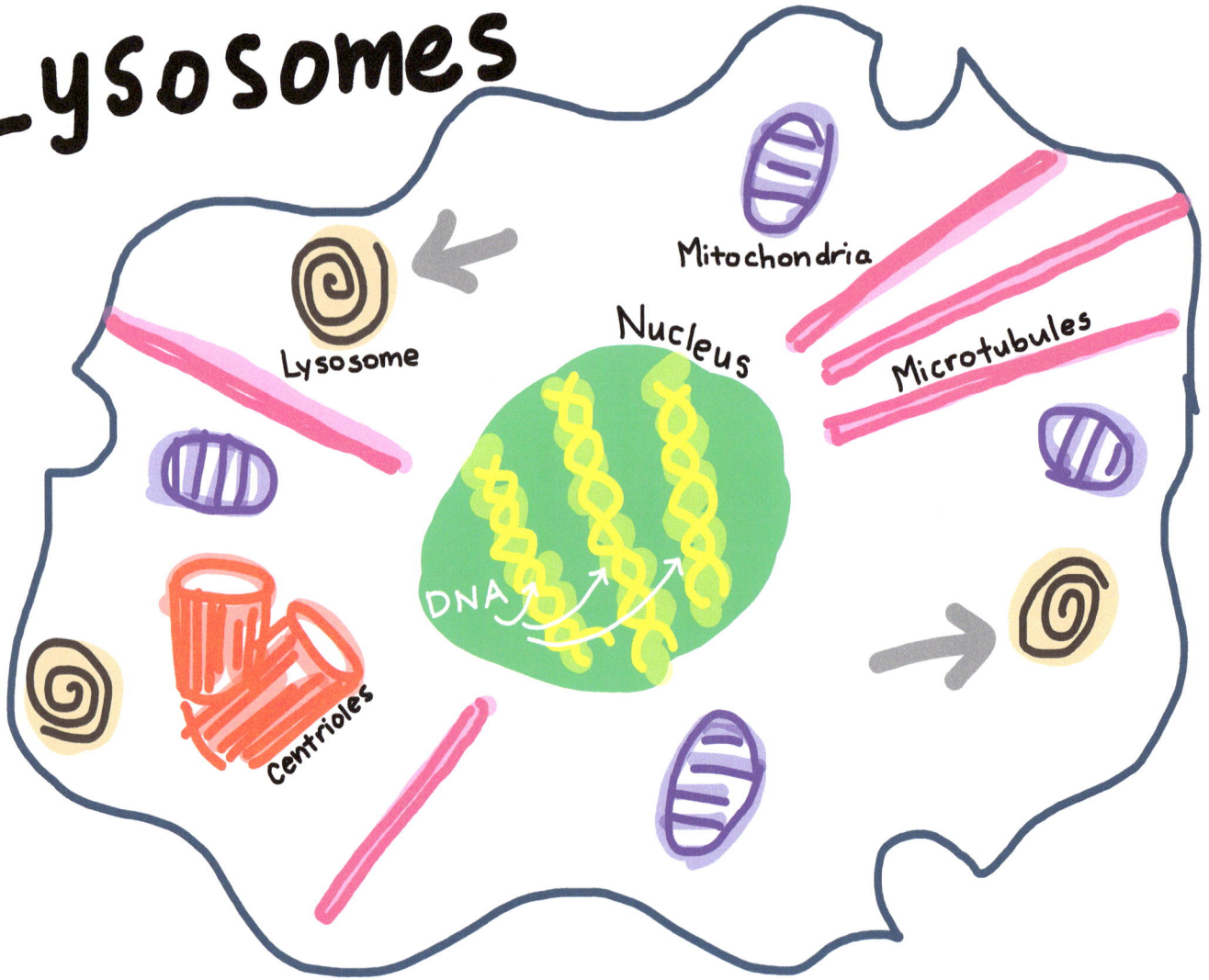

These are **lysosomes**.

Ly-so-somes

Sound It Out

1. LI
2. SO
3. SOMZ

Lysosomes are organelles inside the cell.

Lysosomes are the <u>trash cans</u> of the cell.

What is inside the nucleus?

What do the mitochondria do inside the cell?

How many lysosomes do you see?

Point to the centrioles.

Hormones

Lysosome

Mitochondria

Nucleus

Microtubules

DNA

Centrioles

Hormones

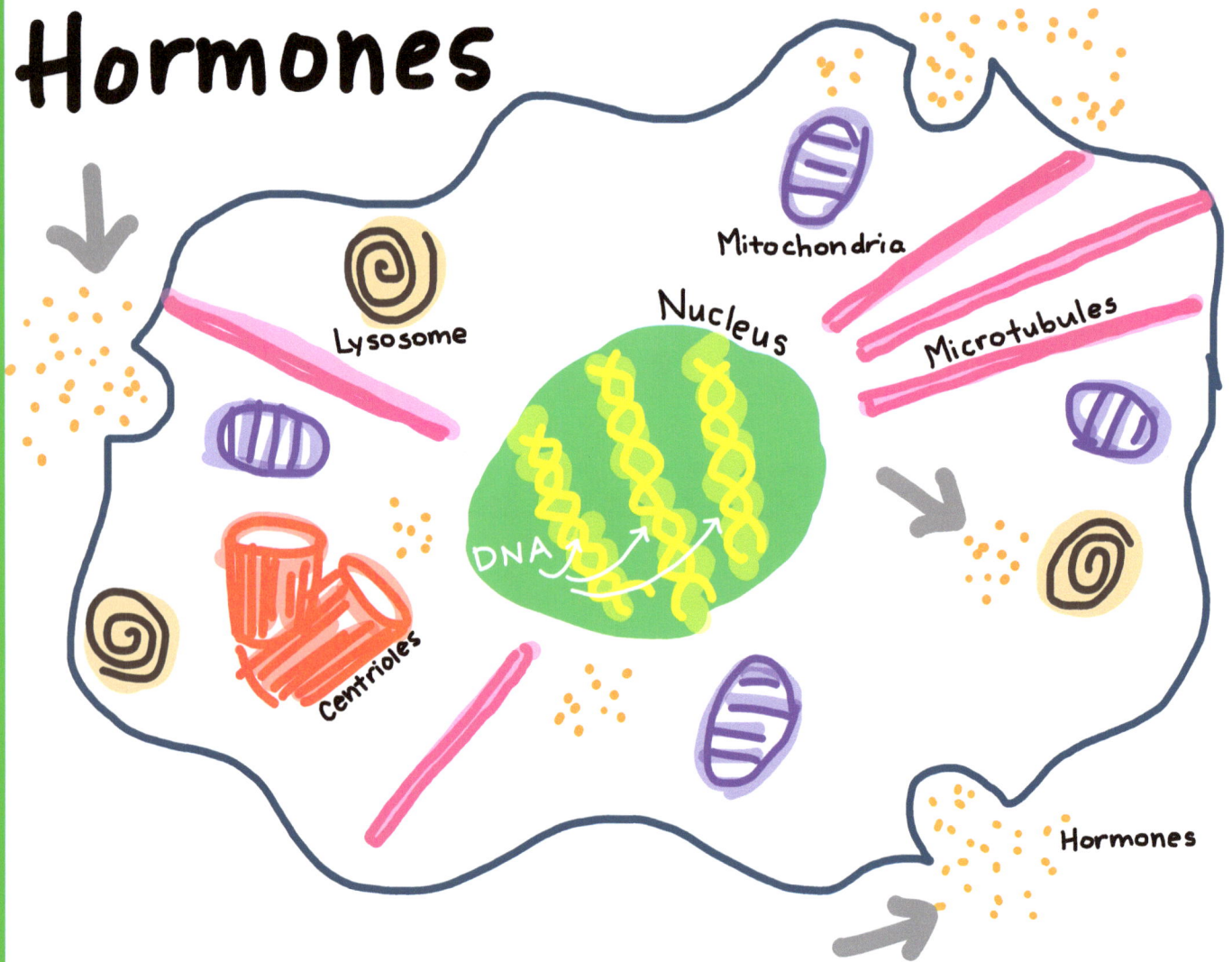

These are hormones.

Hor-mones

Hormones can be inside or outside the cell.

Hormones <u>make you happy</u>!

Do you see the hormones?

Do you see the nucleus with DNA inside?

How many microtubules do you see?

What do the lysosomes do inside the cell?

You have learned
six parts of the cell...

The Cell

Can you name all six?

Golgi

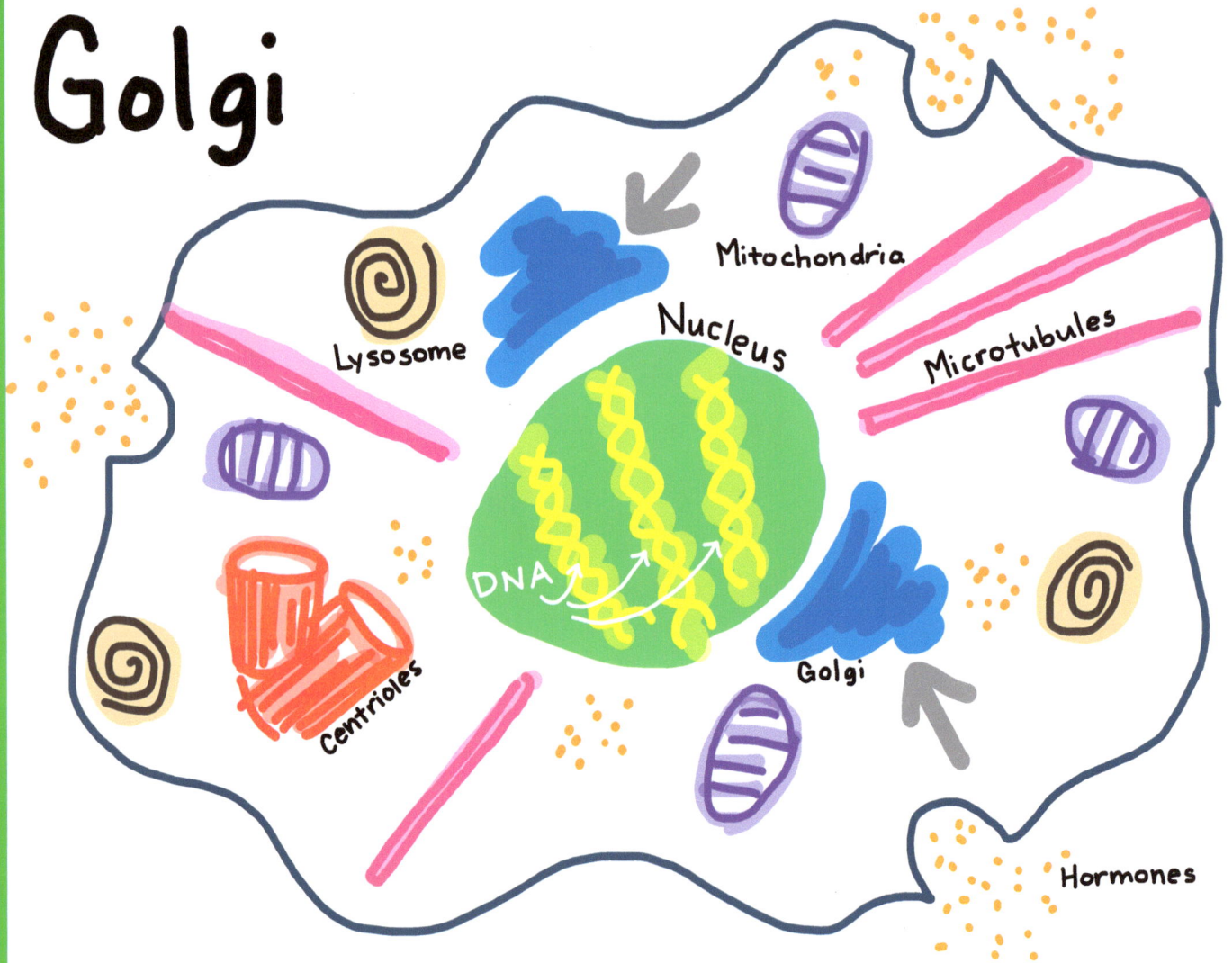

This is the **golgi**.

Gol-gi

The golgi is an organelle inside the cell.

The golgi is made of layers, like pancakes.

The golgi makes and sends packages to other parts of the cell, or out of the cell.

Do you see the golgi?

What do the mitochondria do inside the cell?

What is inside the nucleus?

Ribosomes

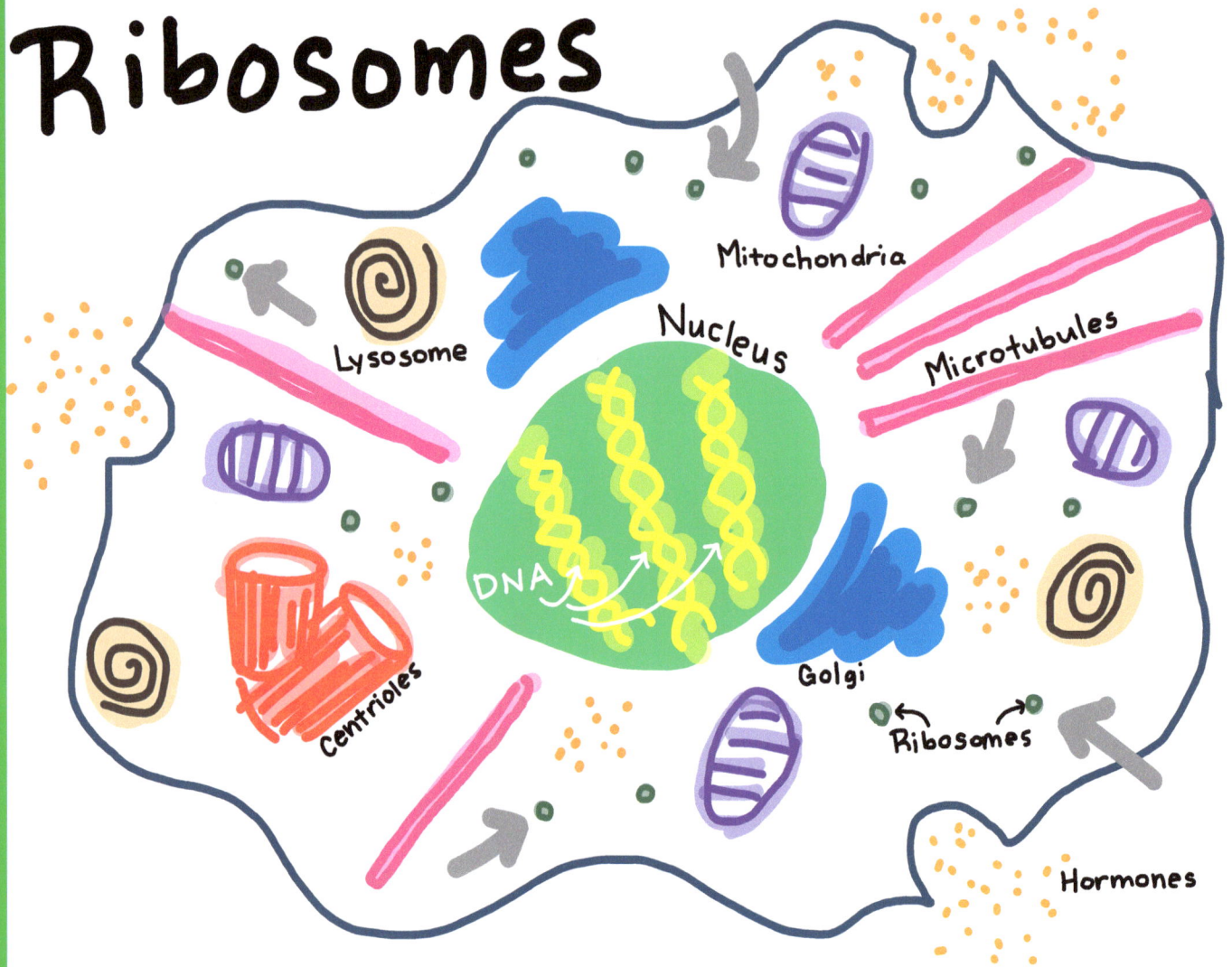

Lysosome
Mitochondria
Nucleus
Microtubules
DNA
Golgi
Centrioles
Ribosomes
Hormones

These are the **ribosomes**.

Ri-bo-somes

Sound It Out
1. RI
2. BO
3. SOMZ

Ribosomes float around inside the cell.

Ribosomes **build proteins**.

Where are the *mitochondria*?

Point to the DNA inside the nucleus.

What does the *golgi* do inside the *cell*?

Do you see the *microtubules*?

Membrane Protein

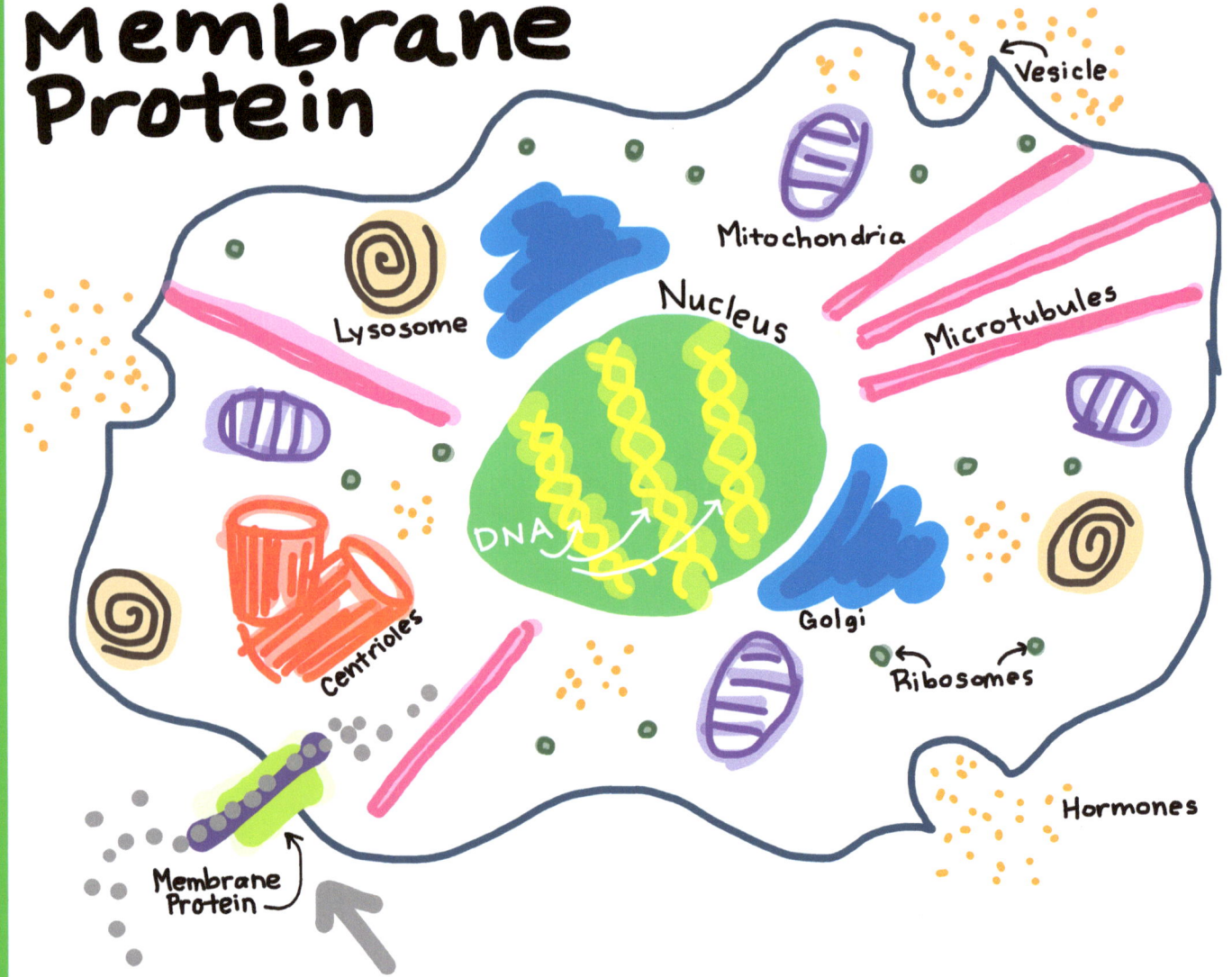

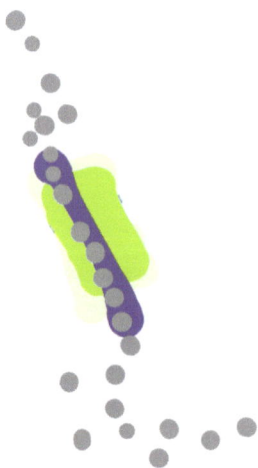

Vesicle

Mitochondria

Lysosome

Nucleus

Microtubules

DNA

Centrioles

Golgi

Ribosomes

Membrane Protein

Hormones

This is a **membrane** **protein**.

Mem-brane Pro-tein

Sound it Out

1. MEM
2. BRAN
3. PRO
4. TEEN

Membrane proteins are stuck in the membrane of the cell.

Membrane proteins move small things in and out of the cell.

Point to the small things moving in and out of the cell.

Where are the centrioles? Do you see the DNA and nucleus?

What do the mitochondria do inside the cell?

Vesicle

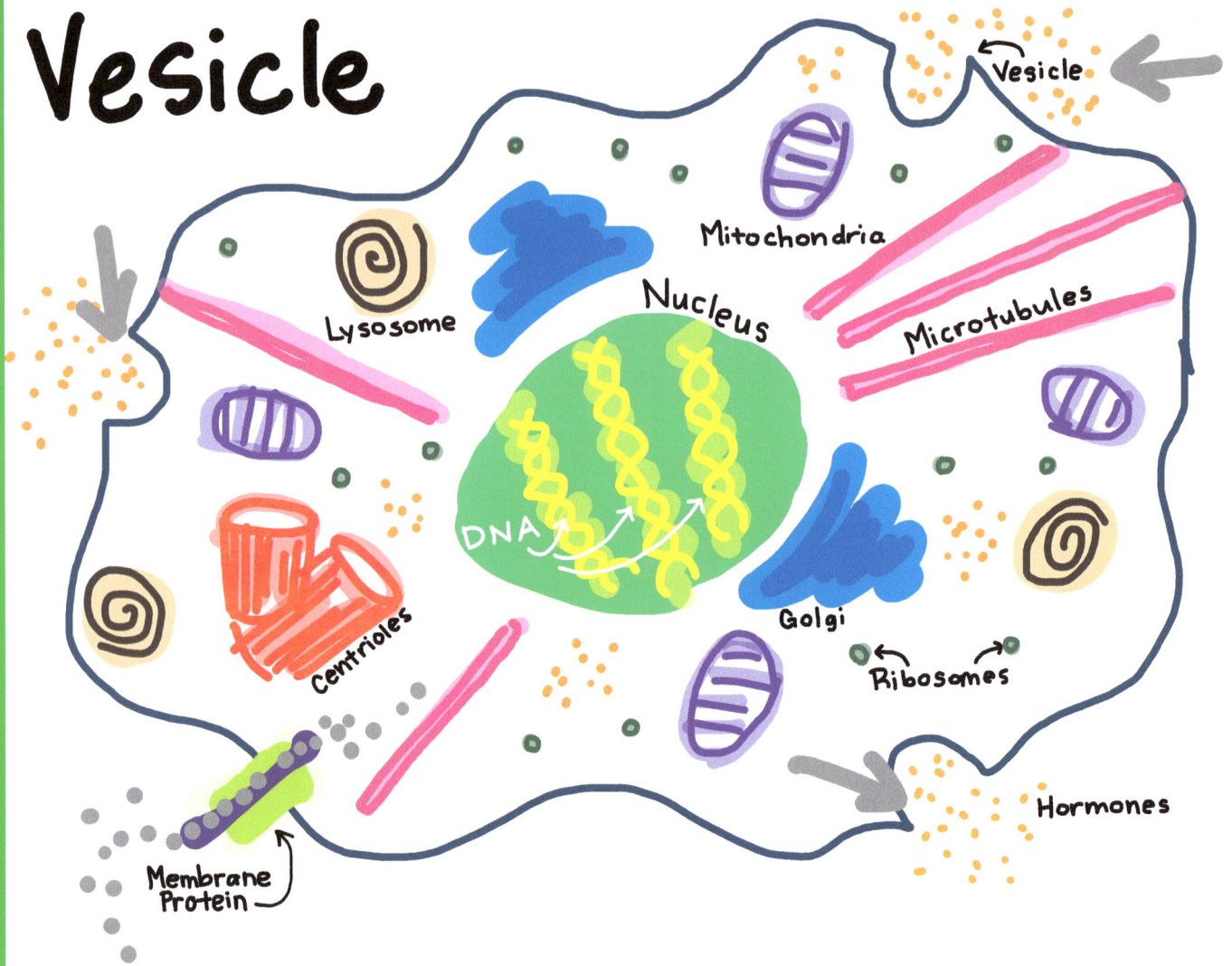

Vesicle

Mitochondria

Lysosome

Nucleus

Microtubules

DNA

Golgi

Ribosomes

Centrioles

Membrane Protein

Hormones

Ves-i-cles

Sound it Out
1. VES
2. EH
3. KLZ

These are vesicles.

Vesicles are <u>little packages</u>. **Vesicles** carry things like **hormones** in and out of the **cell**.

Point to the **vesicles** carrying **hormones** in and out of the **cell**.

Where is the *membrane* *protein*?

What do the *microtubules* *do in the* *cell*?

Where is the *nucleus* *and* *DNA*?

Cytoplasm

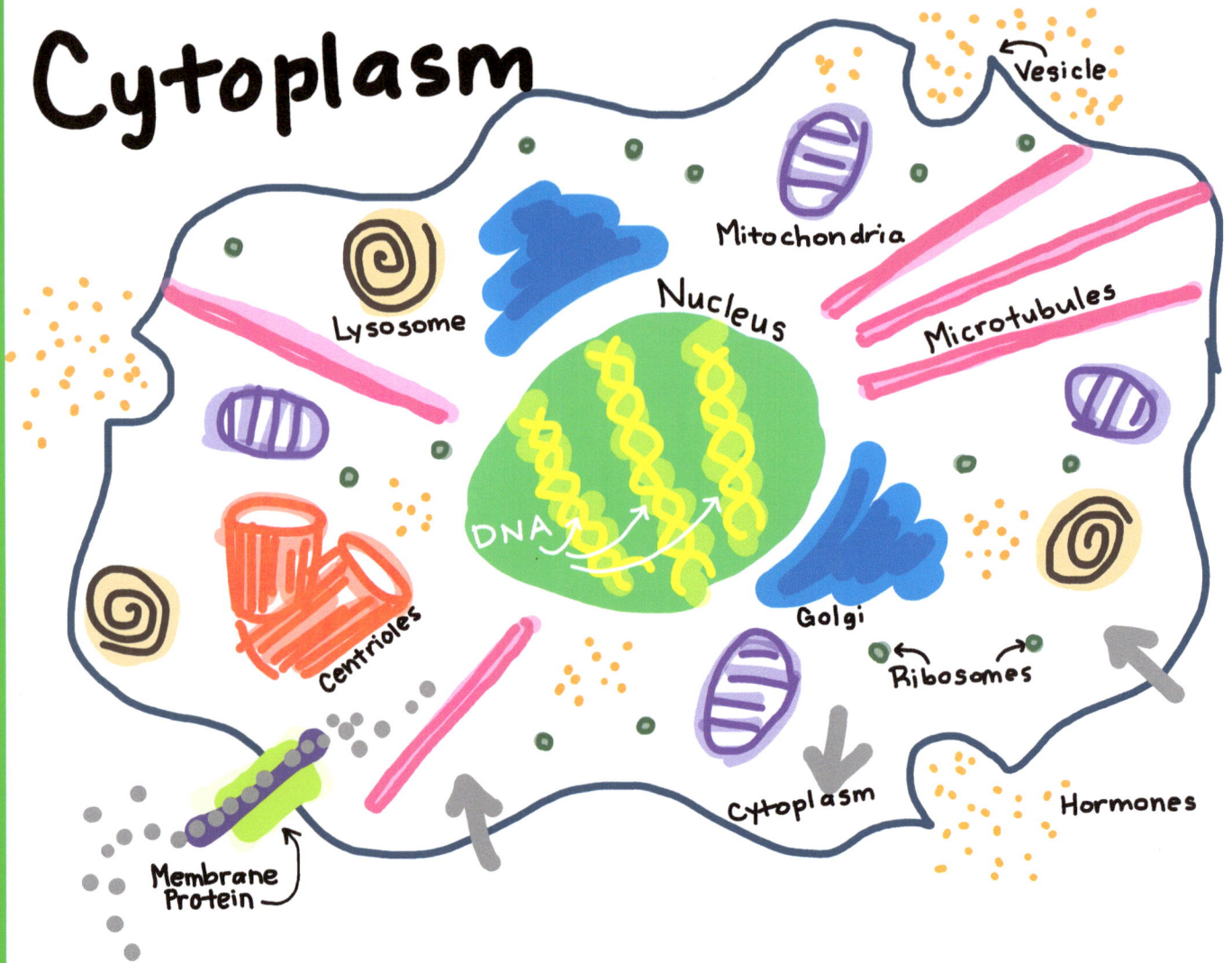

Vesicle

Mitochondria

Nucleus

Microtubules

Lysosome

DNA

Centrioles

Golgi

Ribosomes

Cytoplasm

Hormones

Membrane
Protein

This is the cytoplasm.

Cy-to-plas-m

Sound it Out

1. SI
2. TO
3. PLAZ
4. M

Do you see the open space inside the cell?

This space is filled with cytoplasm.

All of the organelles that you see are floating in cytoplasm inside the cell.

Do you see the centrioles?

Do you know what the golgi does in the cell?

Point to the mitochondria.

Cell Membrane

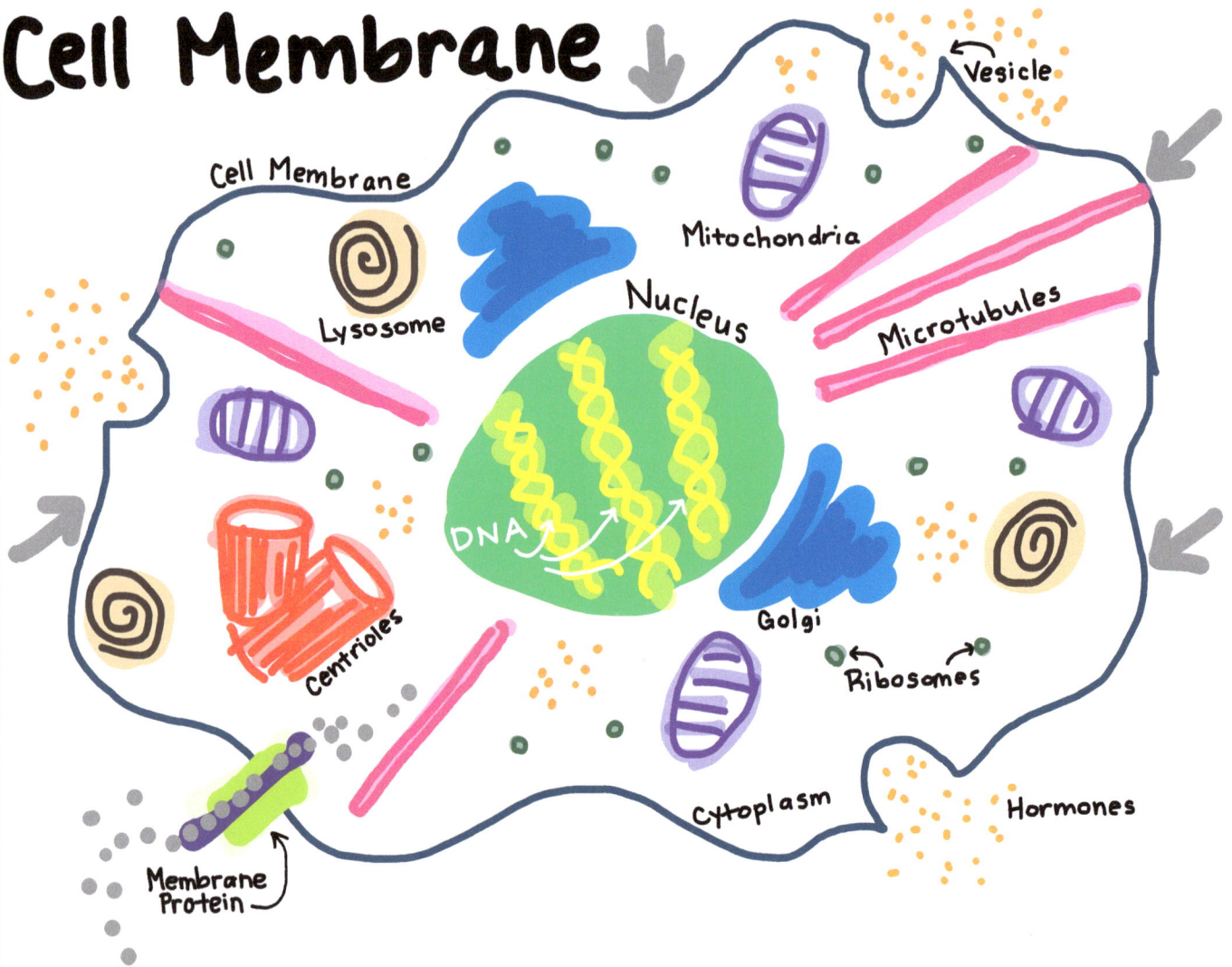

Cell Membrane

Lysosome

Cell Membrane

Nucleus

DNA

Mitochondria

Microtubules

Vesicle

Centrioles

Golgi

Ribosomes

Membrane Protein

Cytoplasm

Hormones

This is the cell membrane.

Cell Mem-brane

Sound it Out

1. SEL
2. MEM
3. BRAN

The cell membrane keeps all of the organelles and cytoplasm together.

Look at everything inside the cell membrane.

Can you point to the centrioles, nucleus, ribosomes, golgi, hormones, mitochondria, DNA, microtubules, lysosomes, vesicles, membrane protein?

Do you remember
what each organelle does inside the cell?
How many parts of the cell can you name?

The Cell

You are now a Cellular Biology expert!

Chemistry:

The Atom and Elements

Ages 0-75

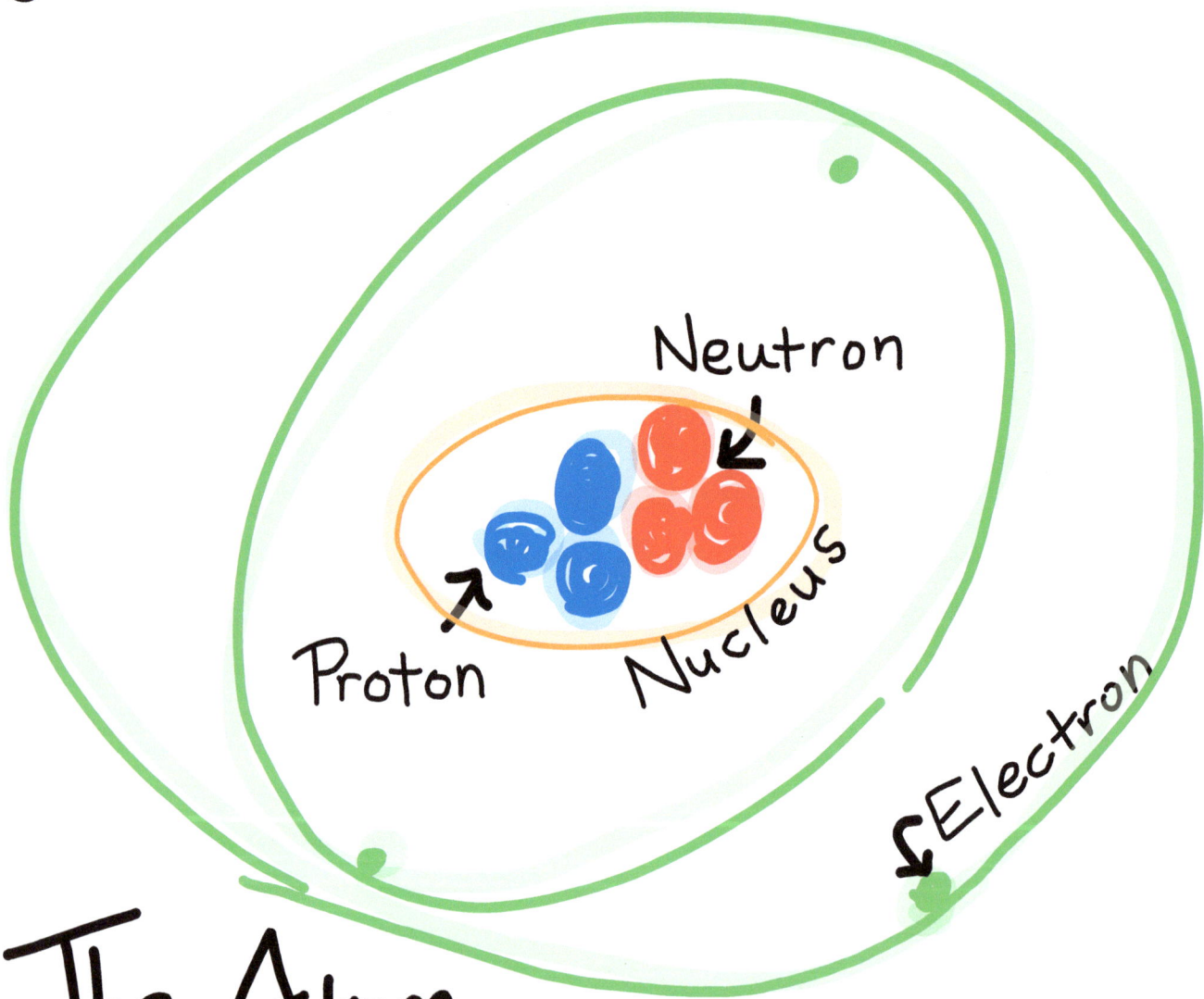

The Atom

Book TWO of the

Super Smart Science Series™

By: APRIL CHLOE TERRAZAS

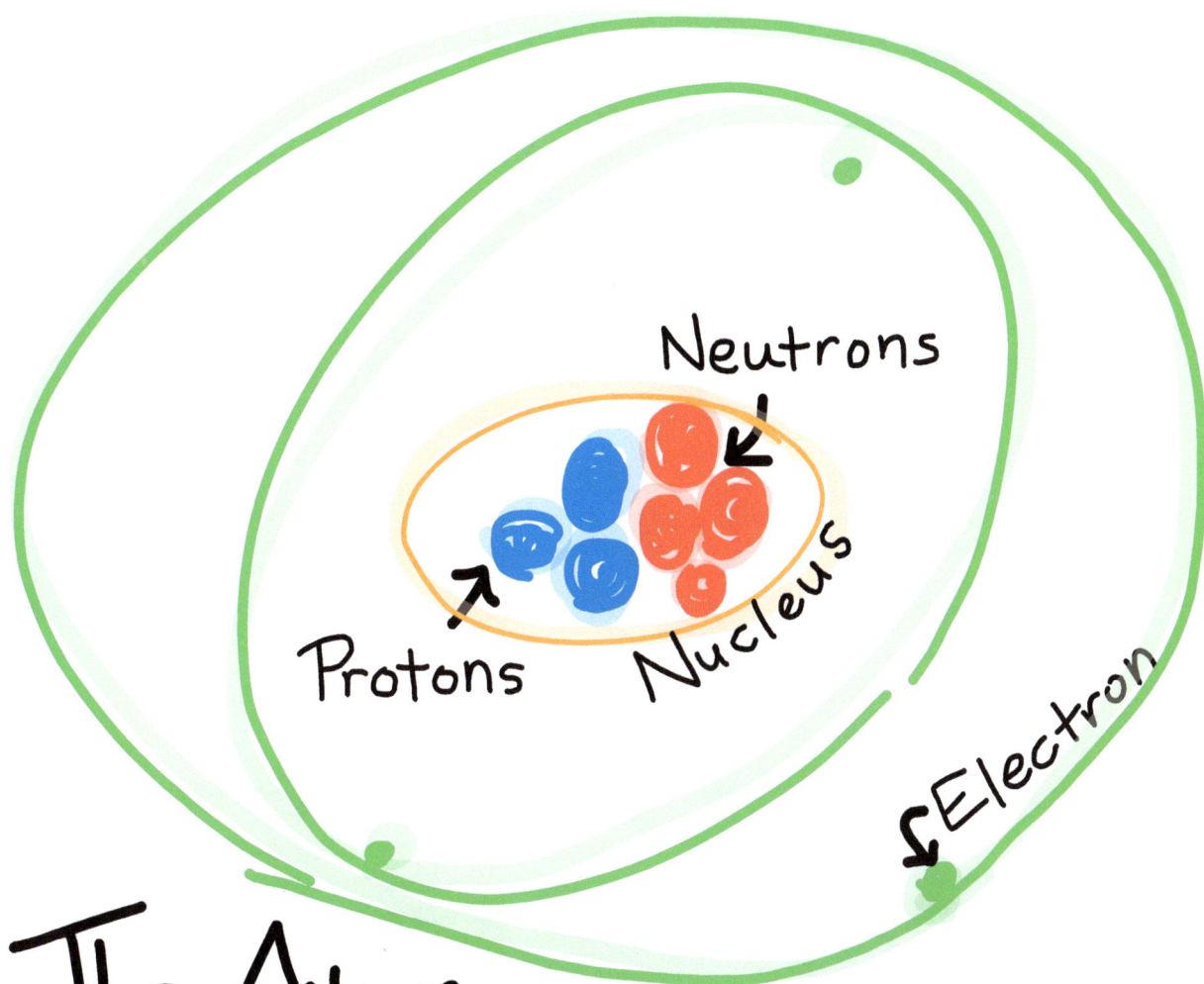

Neutrons

Protons

Nucleus

Electron

The Atom

Chemistry: The Atom and Elements

Written and Illustrated by: APRIL CHLOE TERRAZAS

dedicated to:

DR. IVERSON

OChem Professor Extraordinaire!!!

Thank you for making science FUN & exciting!

Chemistry: The Atom and Elements. April Chloe Terrazas, BS University of Texas at Austin.
Copyright © 2013 Crazy Brainz, LLC

Visit us on the web! www.Crazy-Brainz.com

Cover design, illustrations and text by: April Chloe Terrazas

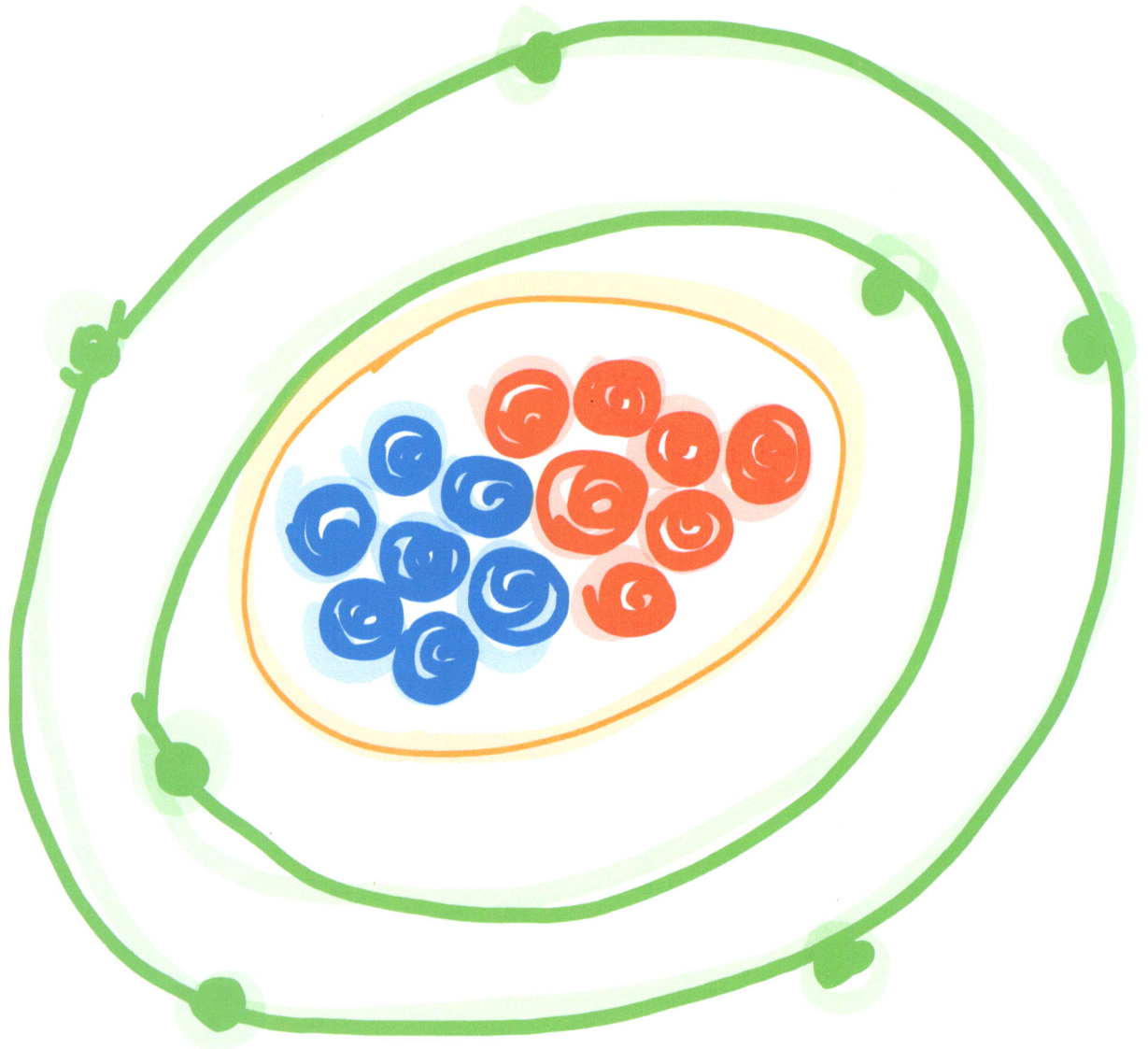

A-tom

El-e-ment

This is an atom
of the element N.

Atoms contain
three different things.

Do you see the blue dots?
The red dots?
And the green dots?

The following pages will
teach you about atoms
and elements.

You are becoming a
Chemistry expert!

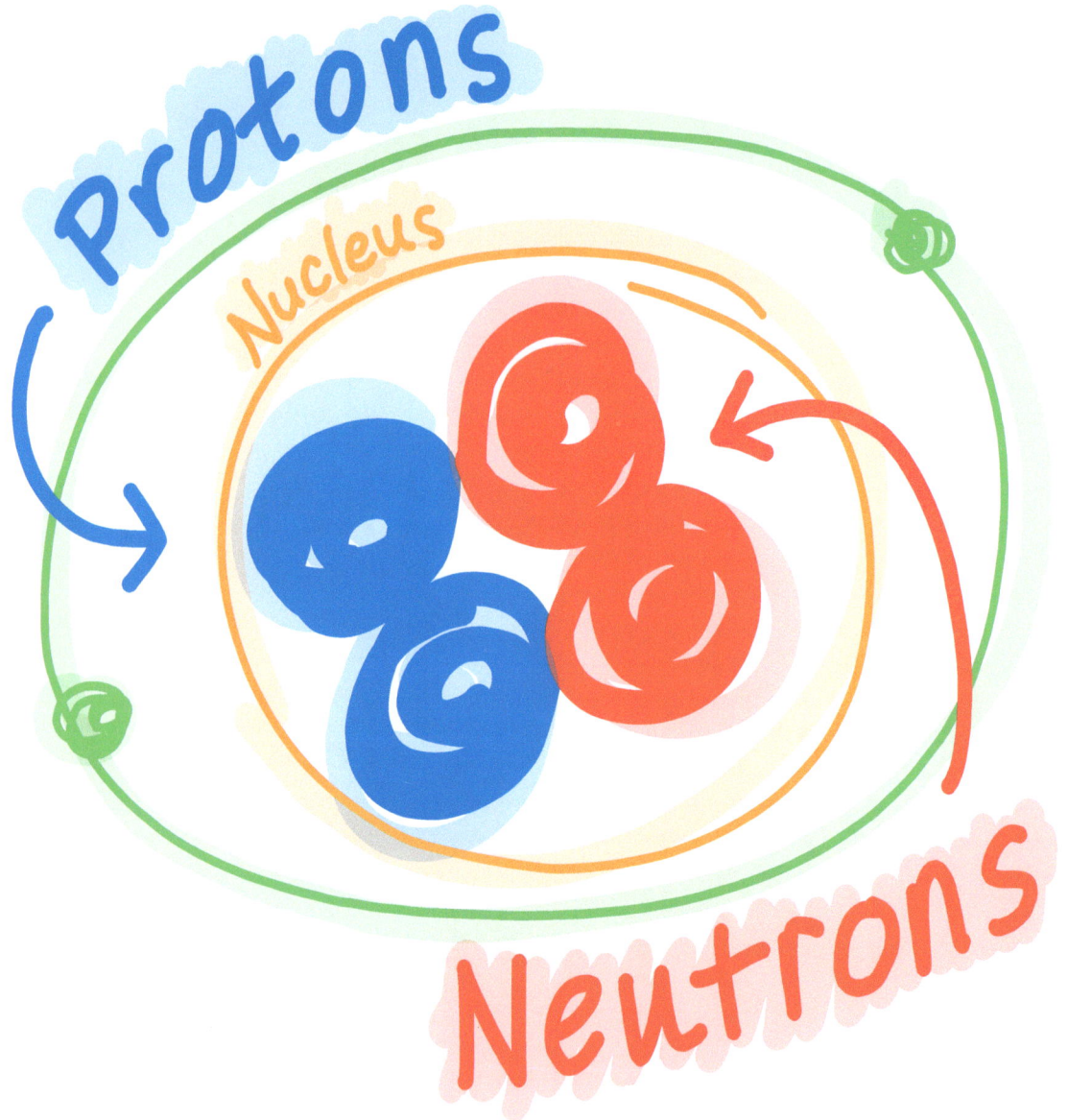

Protons

Nucleus

Neutrons

Pro-ton

Sound it Out
1. PRO
2. TON

Neu-tron

Sound it Out
1. NU
2. TRON

Nu-cle-us

Sound it Out
1. NU
2. KLEE
3. US

Protons and **neutrons** are in the center of the **atom**.

The center of the **atom** is called the **nucleus**.

Different **elements** have different numbers of **protons** and **neutrons** in the **nucleus**.

What are the **green dots** around the **nucleus**?

Turn the page to find out!

Electrons

E-lec-tron

Sound it Out

1. EE
2. LEK
3. TRON

Electrons are outside the nucleus of the atom.

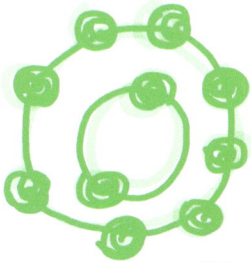

Do you see the electrons outside the nucleus of the atom?

Different elements have different numbers of electrons.

How many electrons are in this atom?

How many protons and neutrons are in this atom?

Periodic Table

1 H	
3 Li	4 Be
11 Na	12 Mg

Per-i-od-ic Ta-ble

Sound it Out

1. PEER
2. EE
3. ODD
4. IK
5. TA
6. BUL

This is the Periodic Table of Elements.

Inside each box is an element.

Of Elements

					2 He
5 B	6 C	7 N	8 O	9 F	10 Ne
13 Al	14 Si	15 P	16 S	17 Cl	18 Ar

Do you see the <u>number</u> next to each element?

This <u>number</u> is called the <u>atomic number</u>.

The <u>atomic number</u> tells you how many protons are inside the nucleus of an atom.

Periodic Table

1 H	
3 Li	4 Be
11 Na	12 Mg

Look at the element C, atomic number 6.

The 6 means that there are 6 protons inside the nucleus of an atom of the element C.

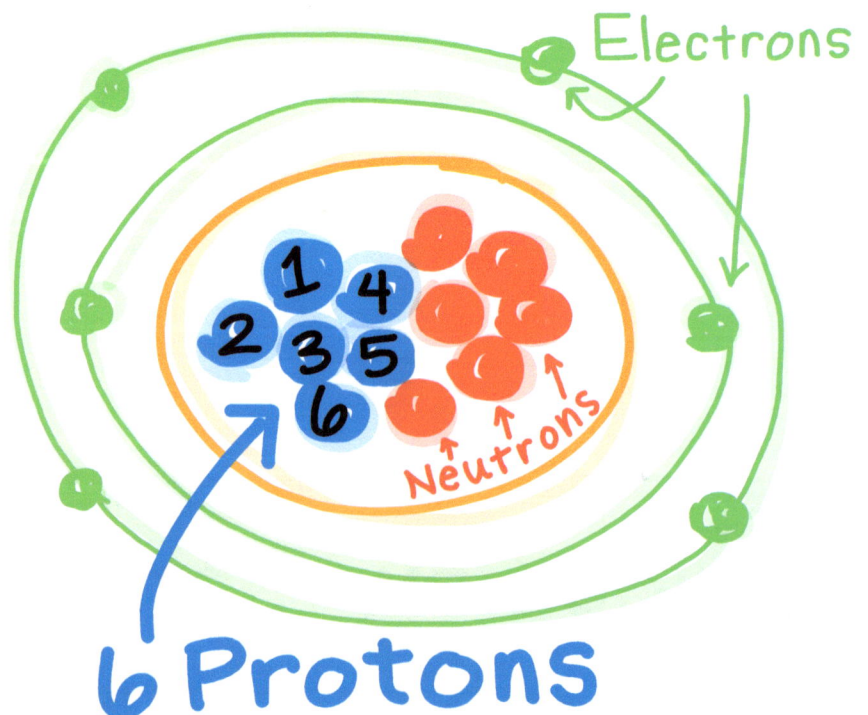

6
C

Electrons

Neutrons

6 Protons

Of Elements

					2 He
5 B	6 C	7 N	8 O	9 F	10 Ne
13 Al	14 Si	15 P	16 S	17 Cl	18 Ar

How many **protons** are inside the **nucleus** of an **atom** of the **element Mg**?

How many **protons** are inside the **nucleus** of an **atom** of the **element Ar**?

Are YOU ready

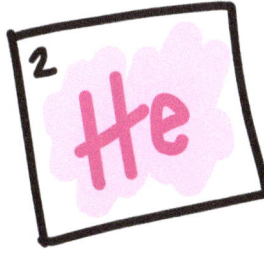

to learn about elements ???

³Li

Lithium

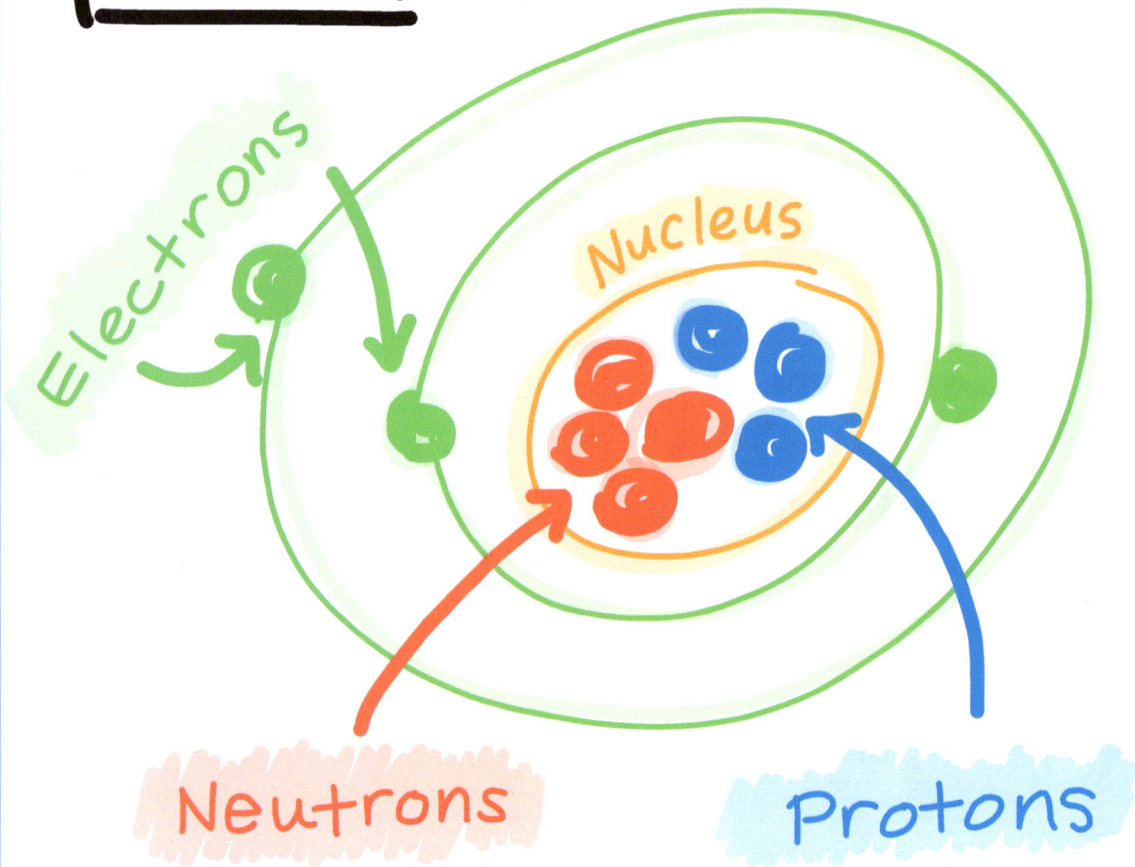

Electrons

Nucleus

Neutrons

Protons

Lith-i-um

Sound it Out
1. **LITH**
2. **EE**
3. **UM**

This is an atom of Lithium.
The symbol for Lithium is Li.

Did you see Li in the
Periodic Table of Elements?

What is the atomic number
of Lithium?

Lithium is in
medicine
and
batteries.

How many protons
are inside the nucleus
of an atom of Lithium?

He Helium

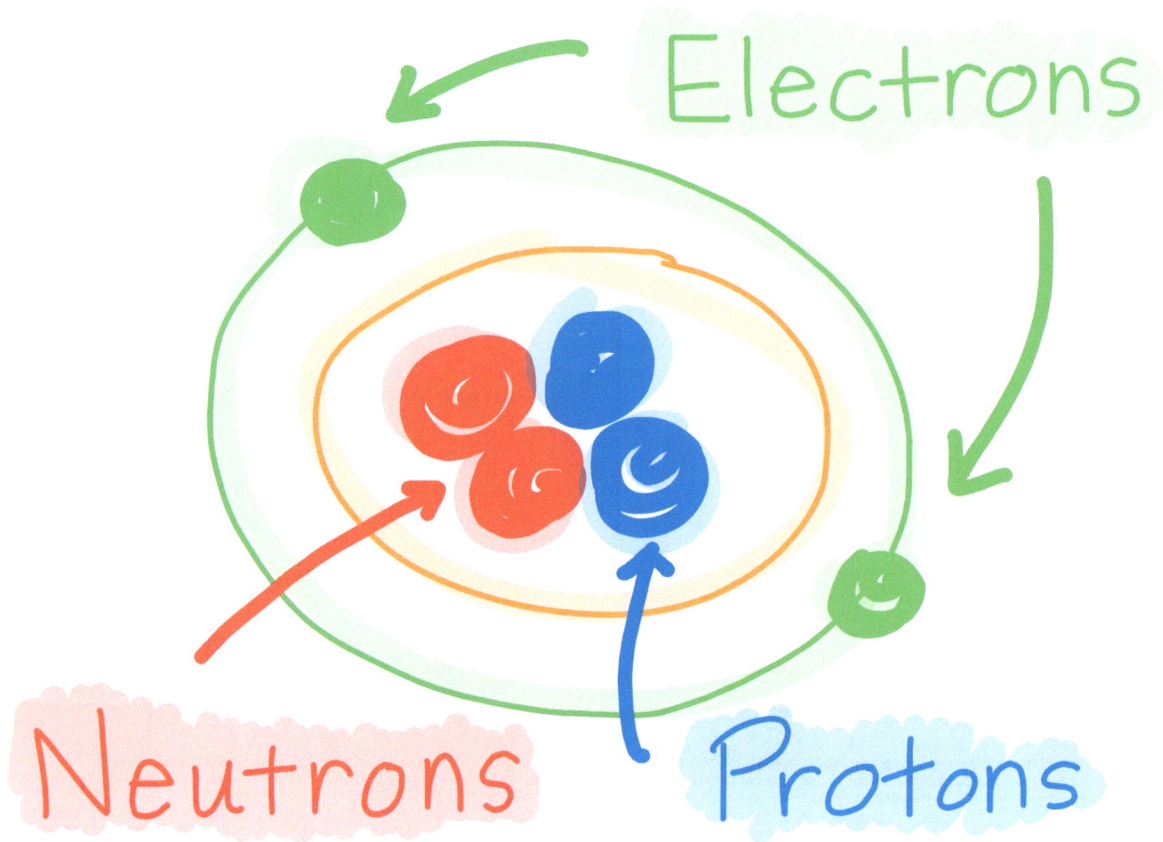

Electrons

Neutrons

Protons

He-li-um

Sound It Out
1. HEE
2. LEE
3. UM

This is an atom of Helium.
The symbol for Helium is He.

Did you see He in the
Periodic Table of Elements?

What is the atomic number
of Helium?

Helium
makes
balloons
float.

How many electrons
are outside the nucleus
of an atom of Helium?

Carbon

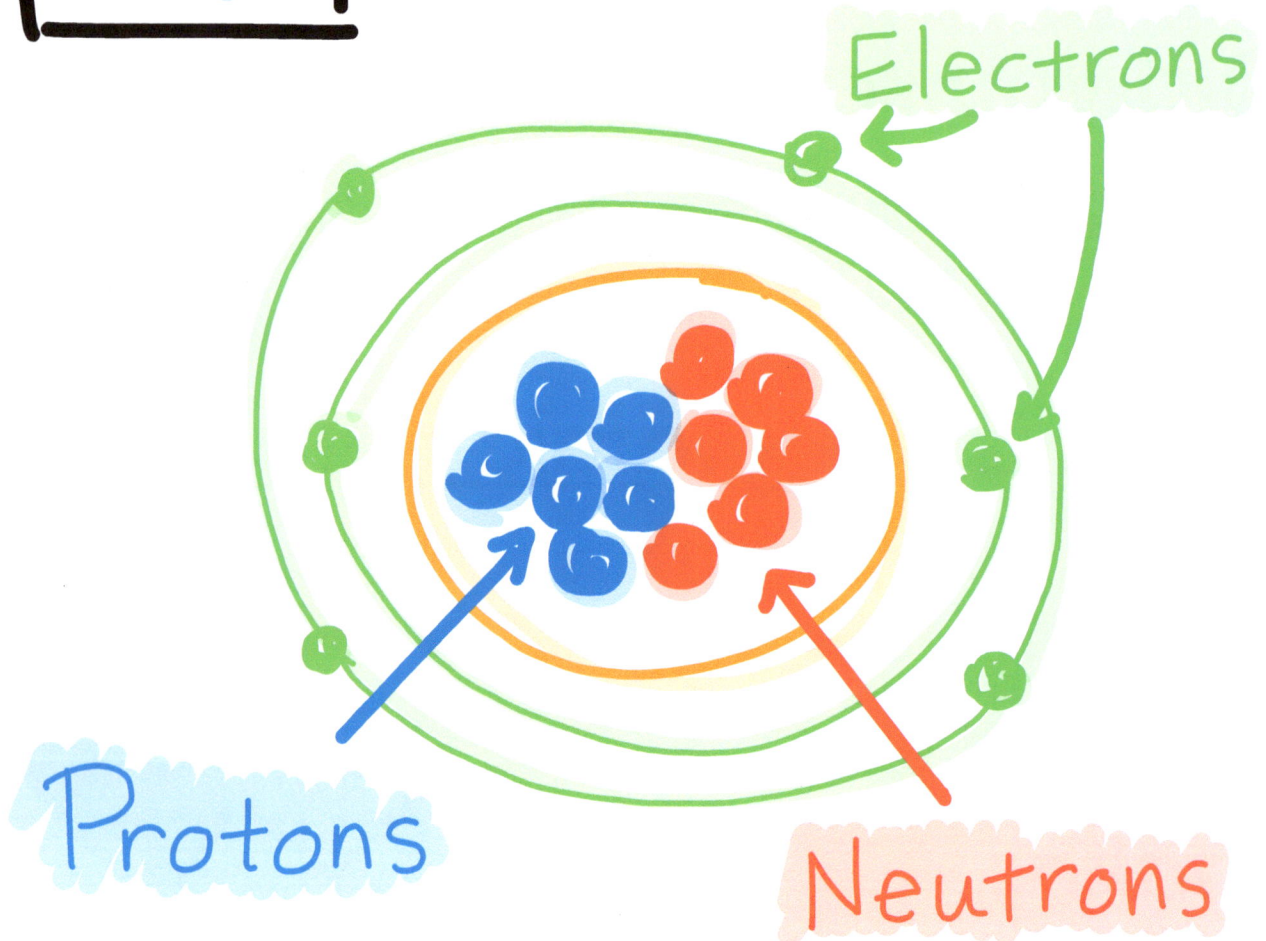

Electrons

Protons

Neutrons

Car-bon

Sound it Out

1. CAR
2. BUN

This is an atom of Carbon.
The symbol for Carbon is C.

Did you see C in the
Periodic Table of Elements?

What is the atomic number
of Carbon?

All living
things are
made of
Carbon.

How many protons
are inside the nucleus
of an atom of Carbon?

You have learned what 3 elements look like as atoms.

Can you name all 3 elements?

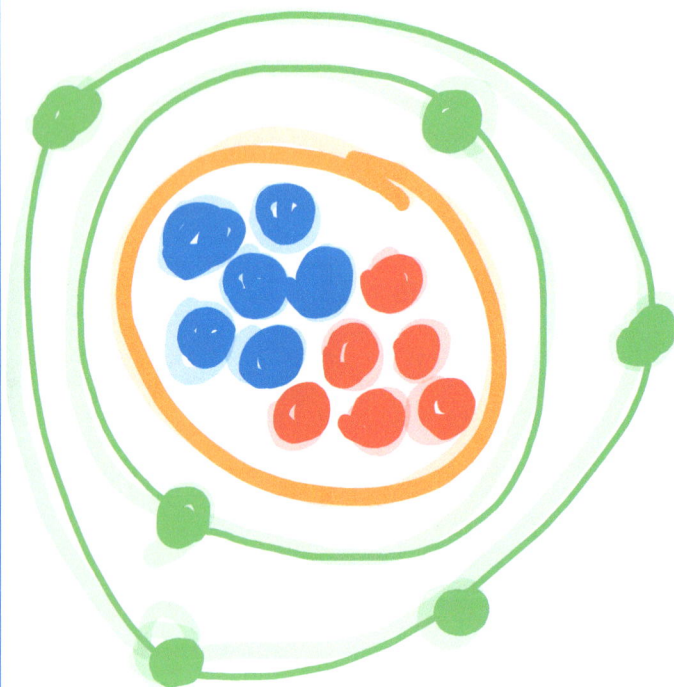

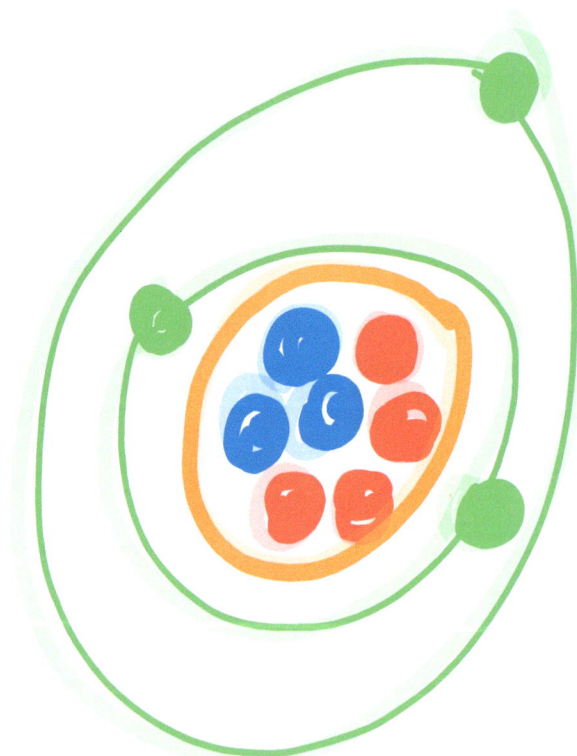

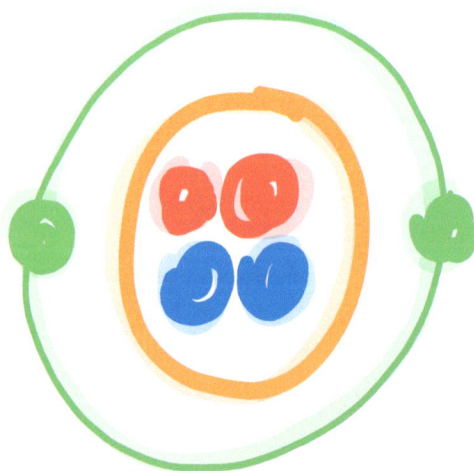

Match the picture with the correct atom.

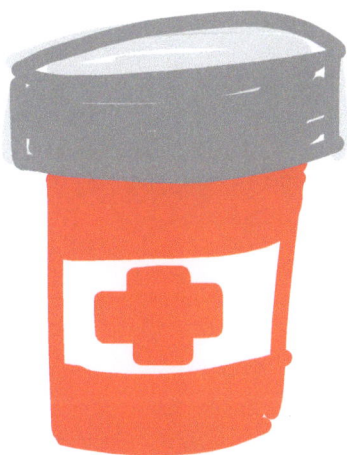

Excellent work! Now we will learn about 3 more elements...

Na Sodium

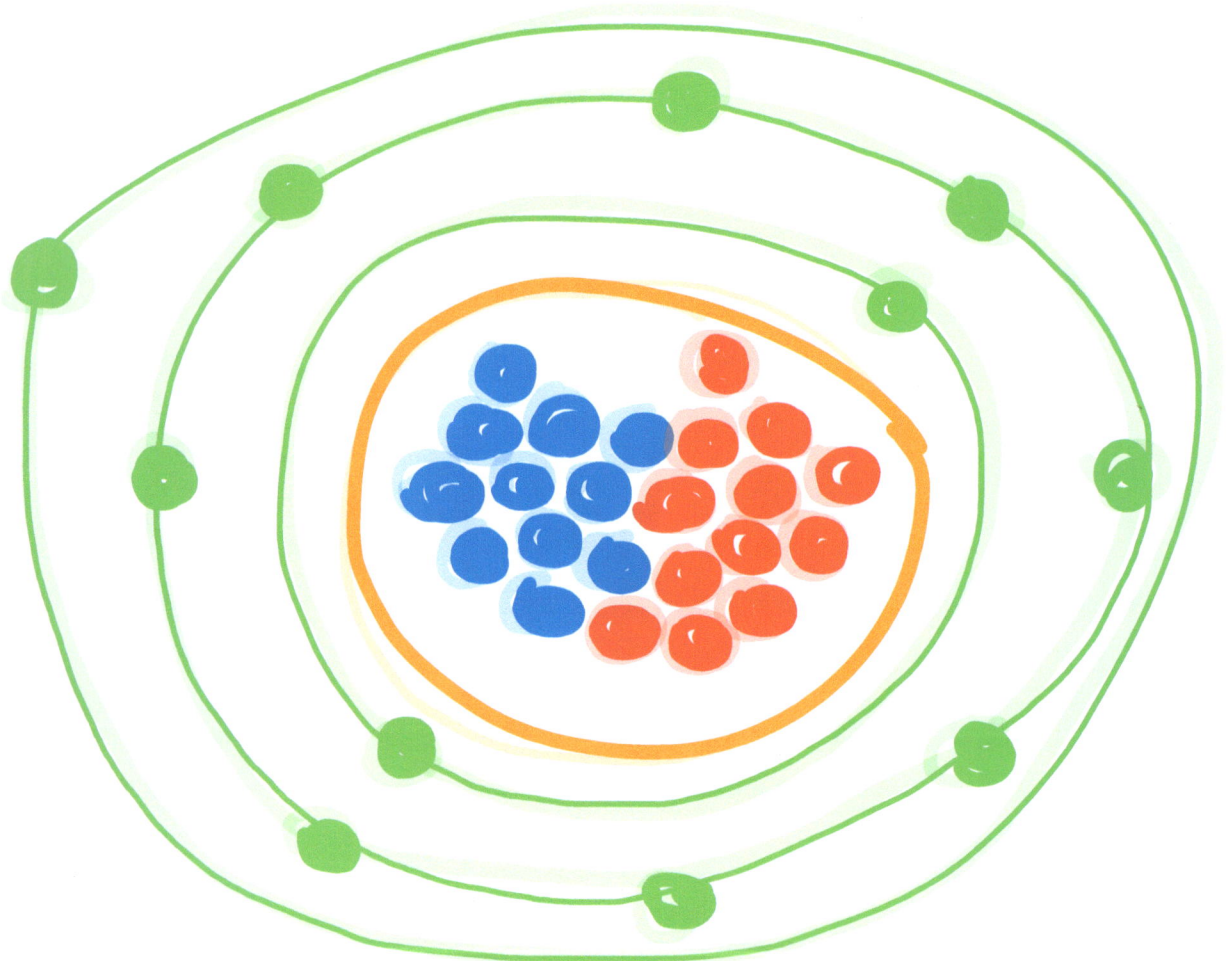

So-di-um

Sound it Out

1. SO
2. DEE
3. UM

This is an atom of Sodium.
The symbol for Sodium is Na.

Did you see Na in the
Periodic Table of Elements?

What is the atomic number
of Sodium?

Sodium
is in
salt.

How many protons
are inside the nucleus
of an atom of Sodium?

10

Ne

Neon

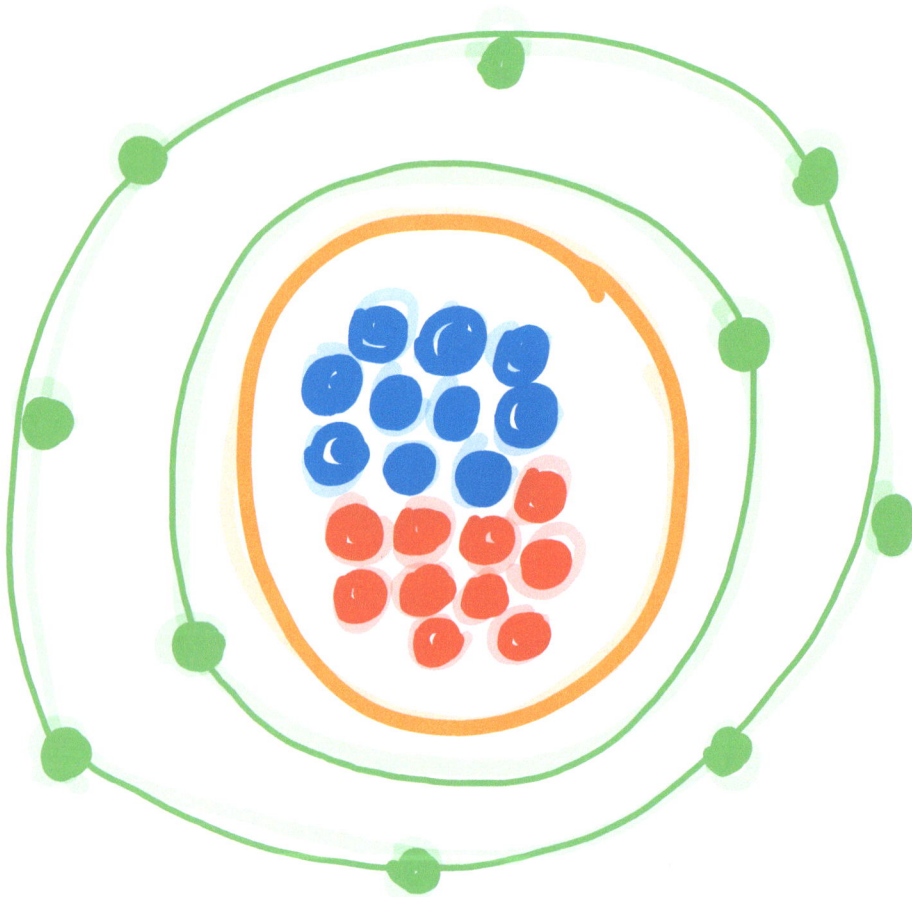

Ne-on

Sound It Out

1. NEE

2. ON

This is an atom of Neon.
The symbol for Neon is Ne.

Did you see Ne in the
Periodic Table of Elements?

What is the atomic number
of Neon?

Neon
is used
to make
signs bright.

OPEN

How many electrons
are outside the nucleus
of an atom of Neon?

O Oxygen

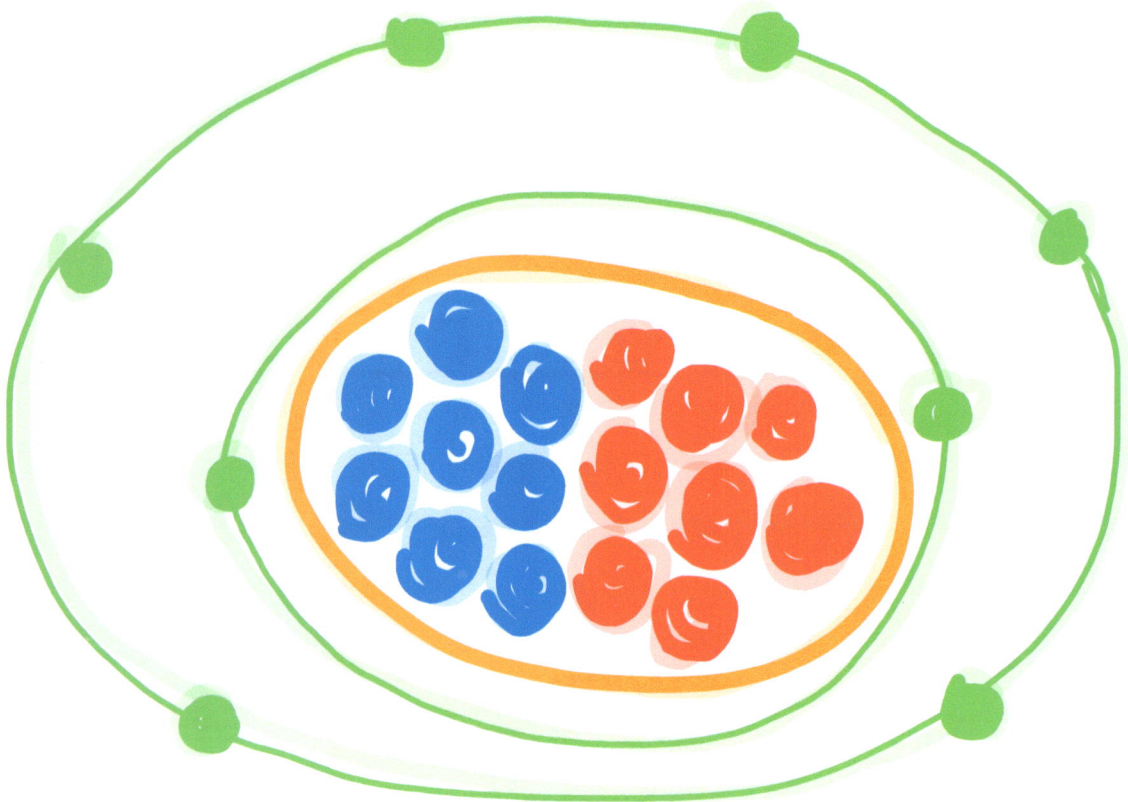

Ox-y-gen

This is an atom of Oxygen.
The symbol for Oxygen is O.

Did you see O in the
Periodic Table of Elements?

What is the atomic number
of Oxygen?

Oxygen is in water.

Water

How many protons
are inside the nucleus
of an atom of Oxygen?

You have learned what
3 more elements look
like as atoms.

Can you name
all 3 elements?

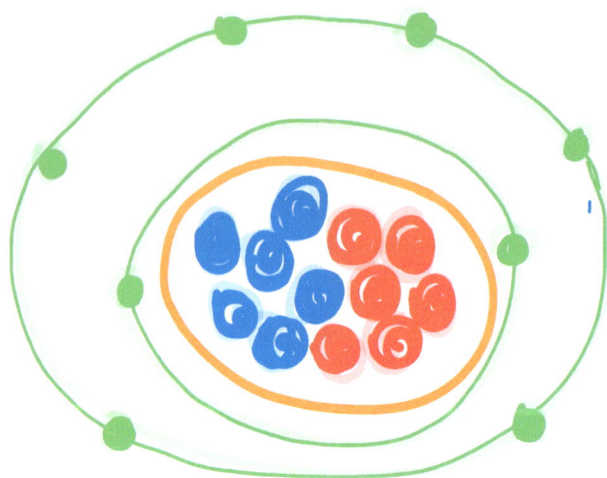

Atomic Number 11

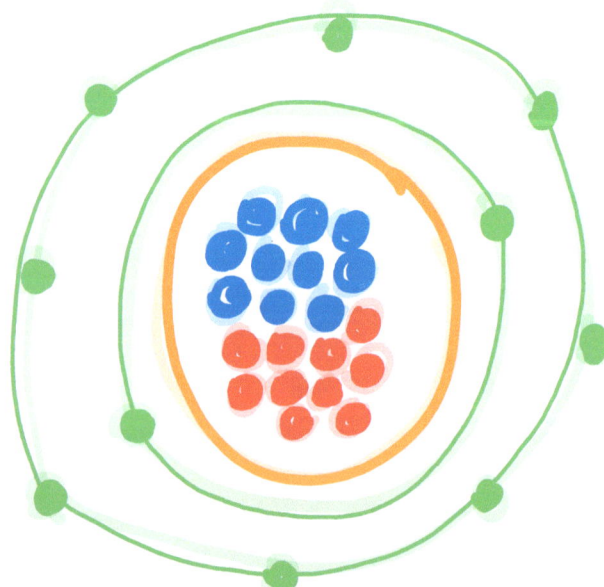

Atomic Number 8

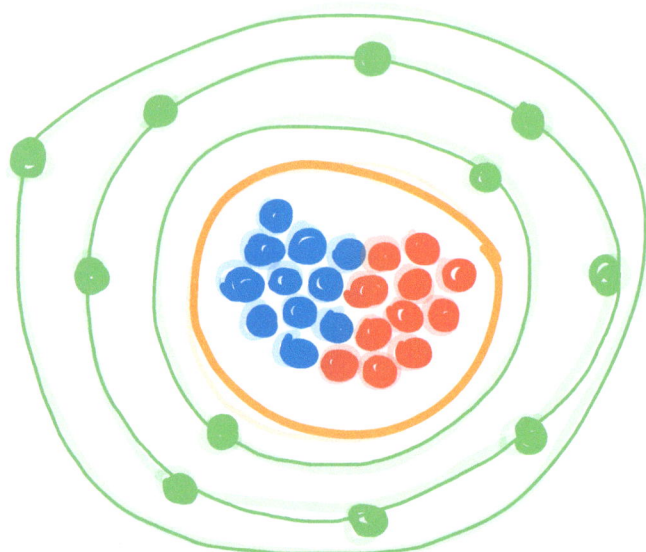

Atomic Number 10

OPEN

Match the picture with the correct atom.

Water

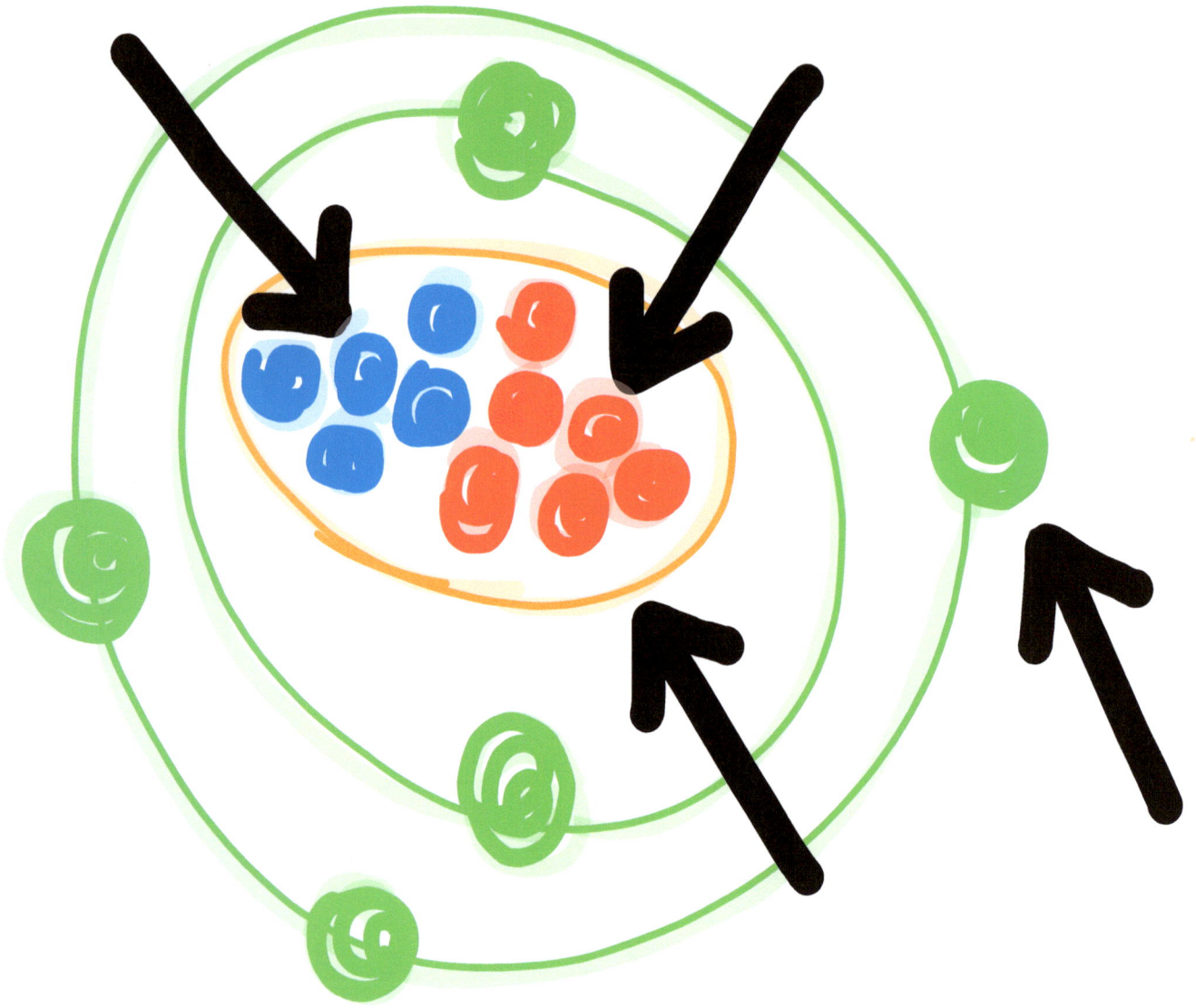

Can you name each part of this atom?

Practice saying the new words you have learned:

Proton

Neutron

Electron

Atom

Lithium

Helium

Carbon

Nucleus

Element

Sodium

Neon

Oxygen

Atomic Number

Periodic Table of Elements

Great job!

You are a Chemistry expert!

NEUROLOGY:

The Amazing Central Nervous System

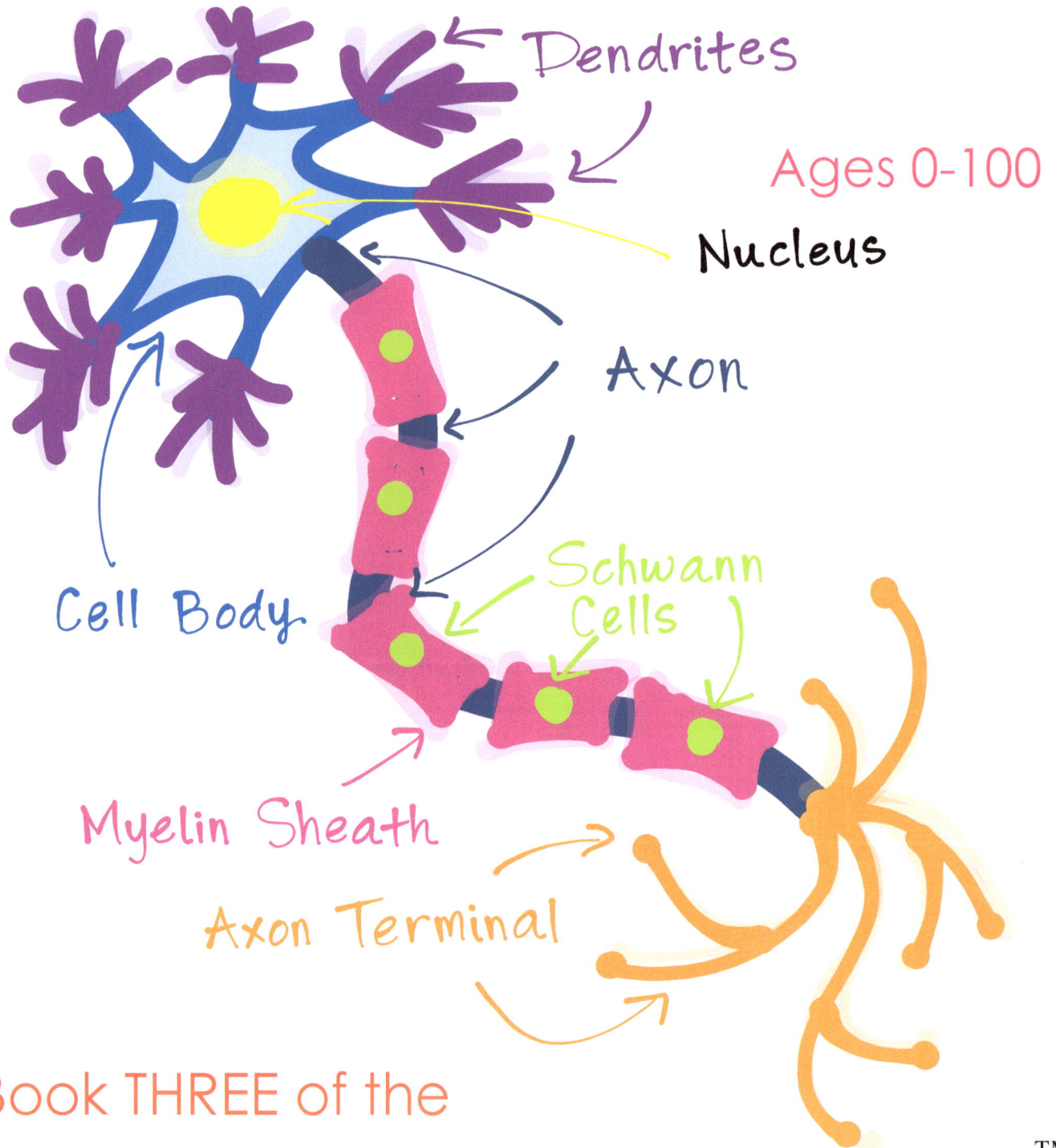

Dendrites

Ages 0-100

Nucleus

Axon

Schwann Cells

Cell Body

Myelin Sheath

Axon Terminal

Book THREE of the

Super Smart Science Series™

BY: APRIL CHLOE TERRAZAS

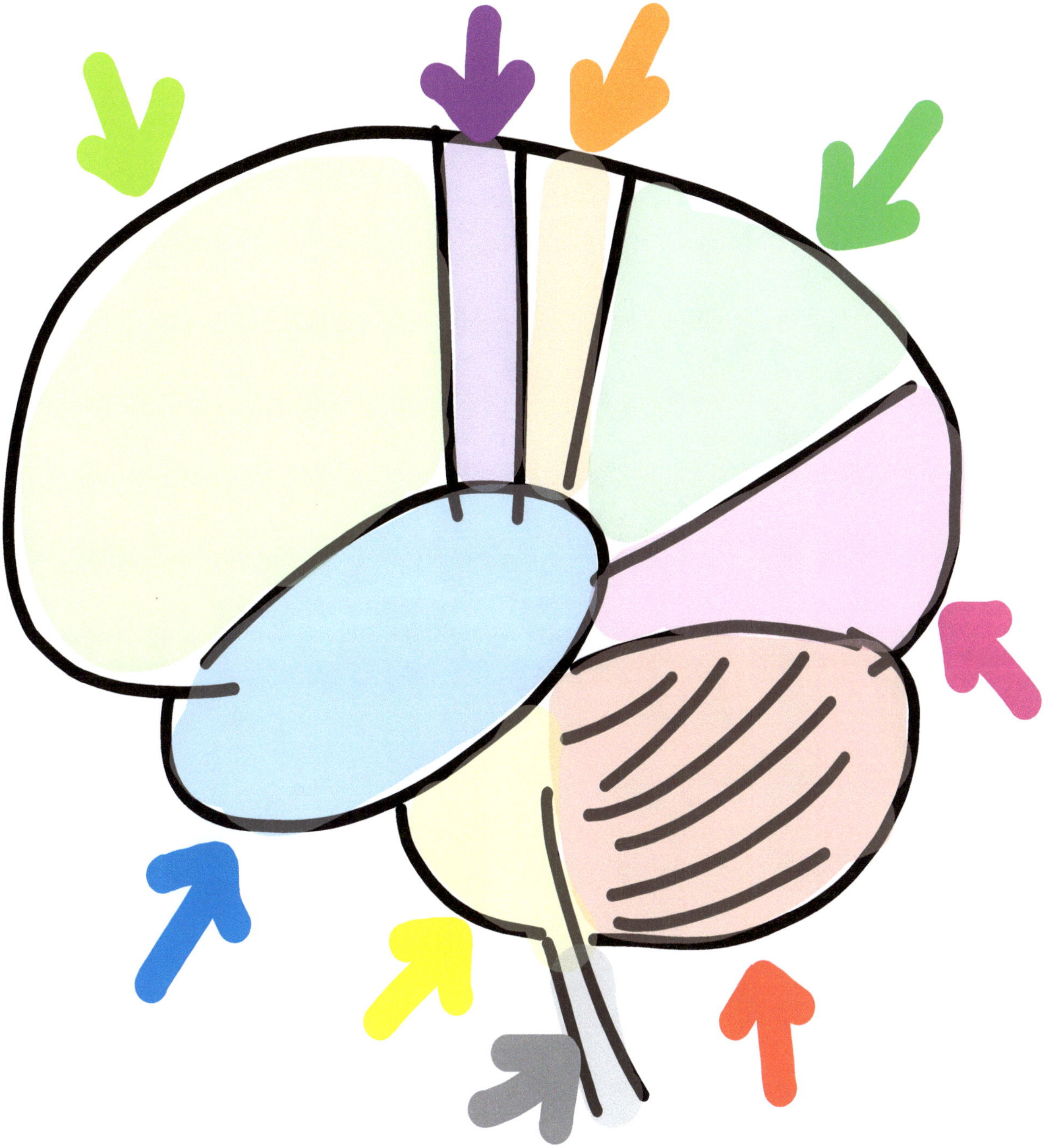

**Winner of the Neurology Art Competition:
John Clint Lawhon!**

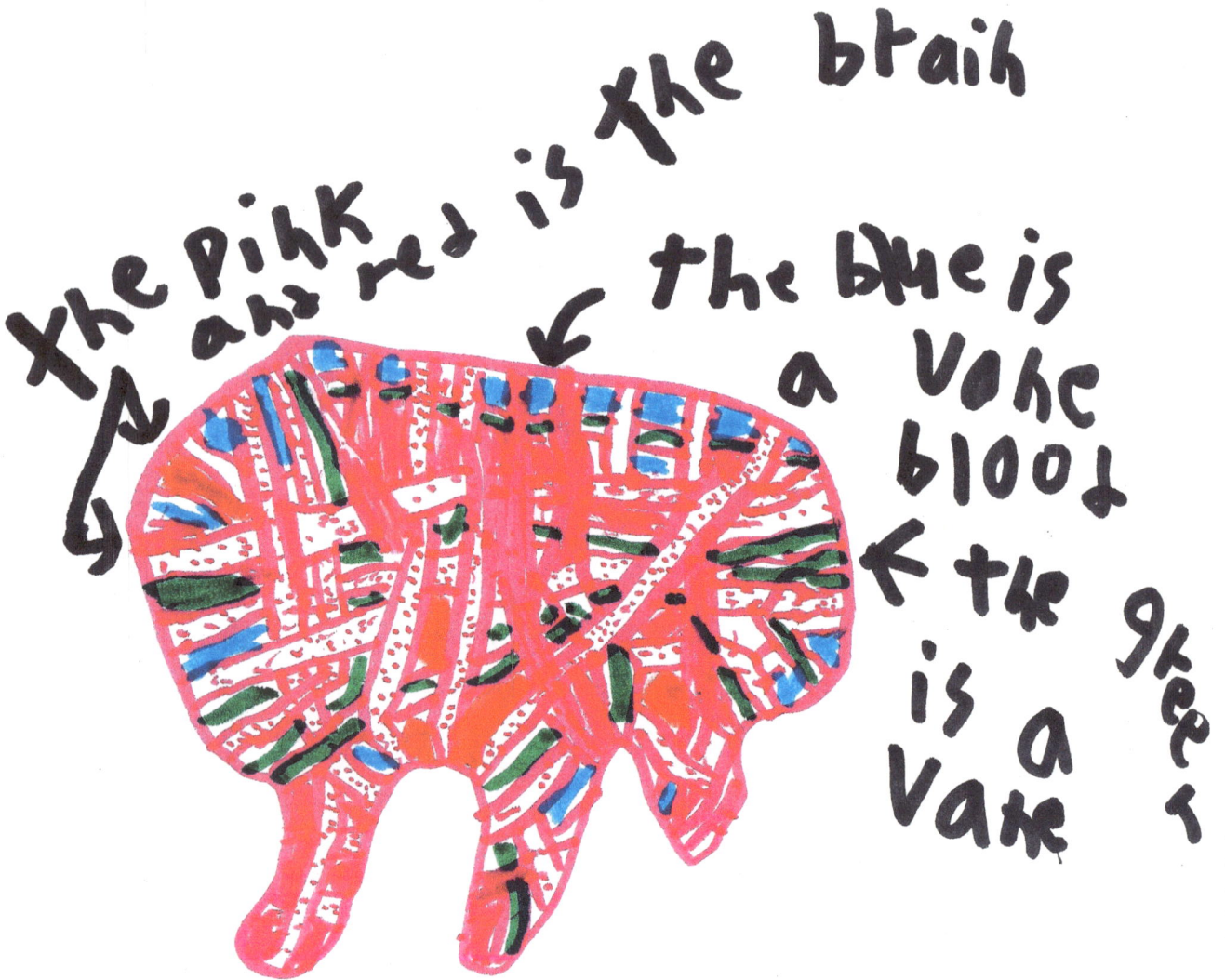

the pink and red is the brain

the blue is a vone bloot

the green is a vane

John Cliht lawhon

Frontal
Lobe

Motor Cortex

Somatosensory Cortex

Parietal
Lobe

Occipital
Lobe

Temporal
Lobe

Brain
Stem

Cerebellum

Spinal Cord

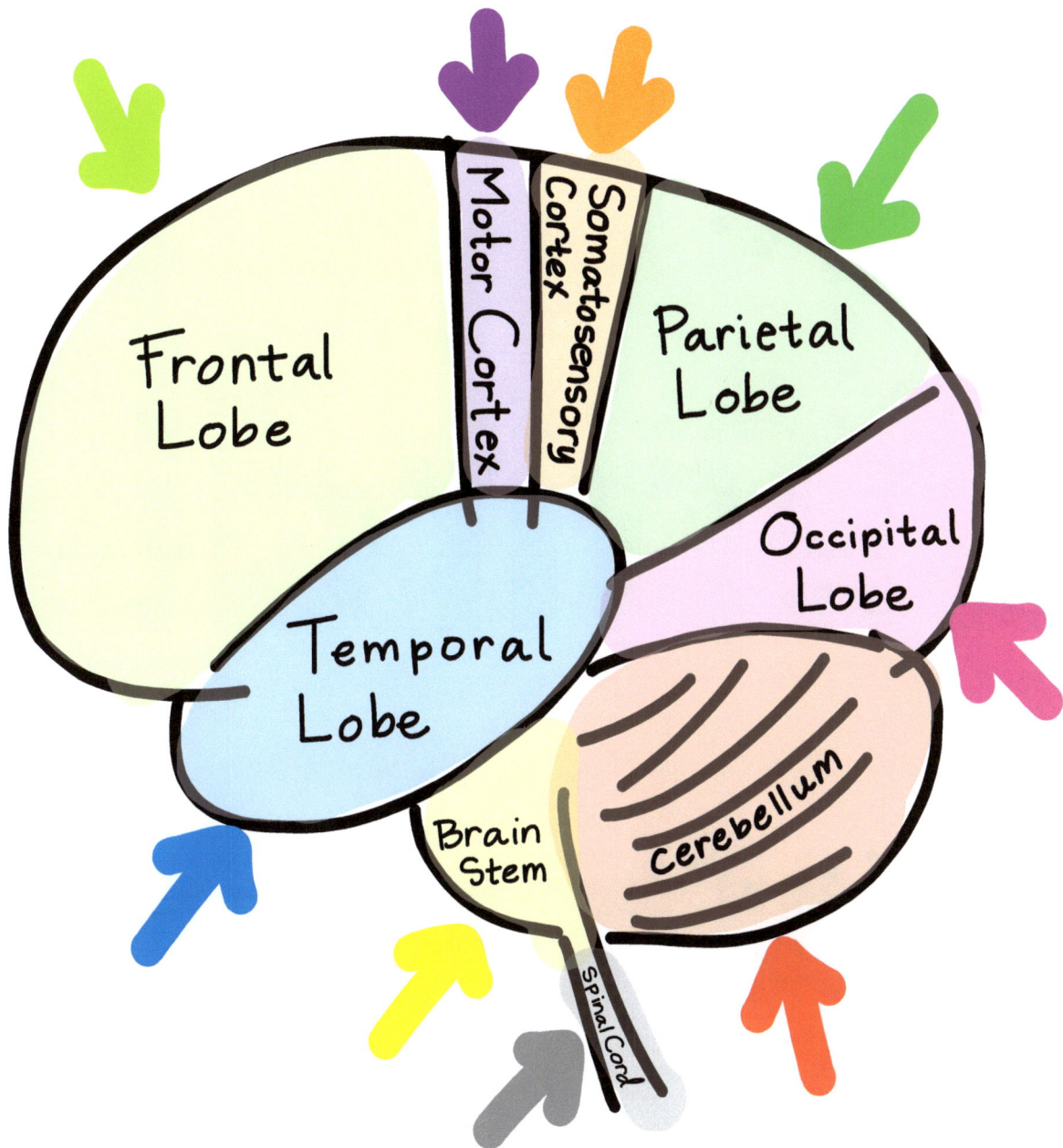

Neurology: The Amazing Central Nervous System

Written and Illustrated by: APRIL CHLOE TERRAZAS

Dedicated to:

Granny Emma Marie

Thank You for your support!!! ☺

Neurology: The Amazing Central Nervous System. April Chloe Terrazas, BS University of Texas at Austin.
Copyright © 2013 Crazy Brainz, LLC

Visit us on the web! www.Crazy-Brainz.com

The Central Nervous System is...

Sen-trul

Ner-vus

Sis-tem

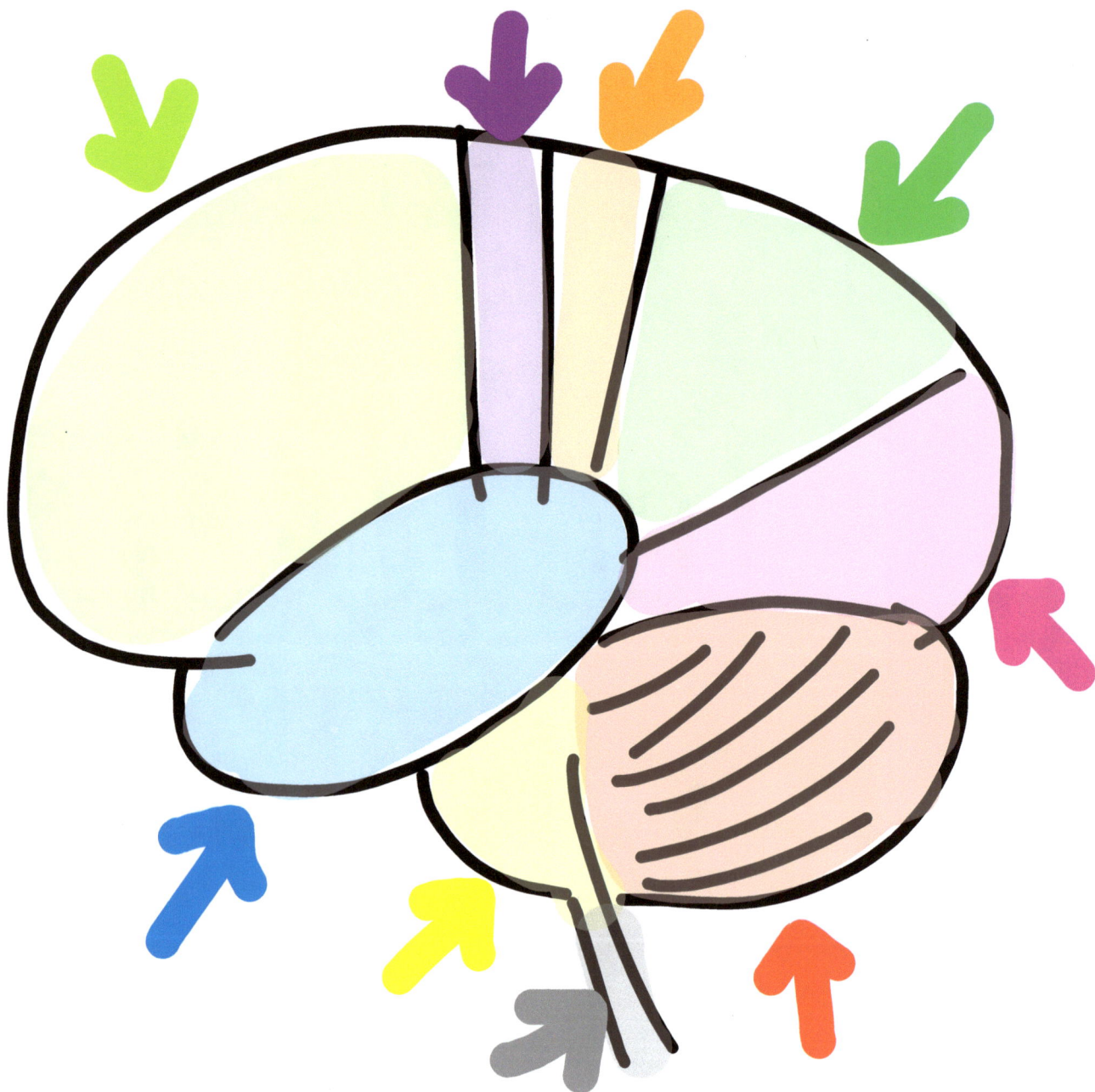

the Brain,

and the Spinal Cord.

The two main organs of the
Central Nervous System
are the **brain** and the **spinal cord**.

The **brain** processes information
received from the **spinal cord**.

First, look at the **brain**.

*Do you see all of the
different colored sections?*

Each part of the **brain**
has a special job.

How many parts of the
brain do you see?

*You are becoming a
Neurology expert!*

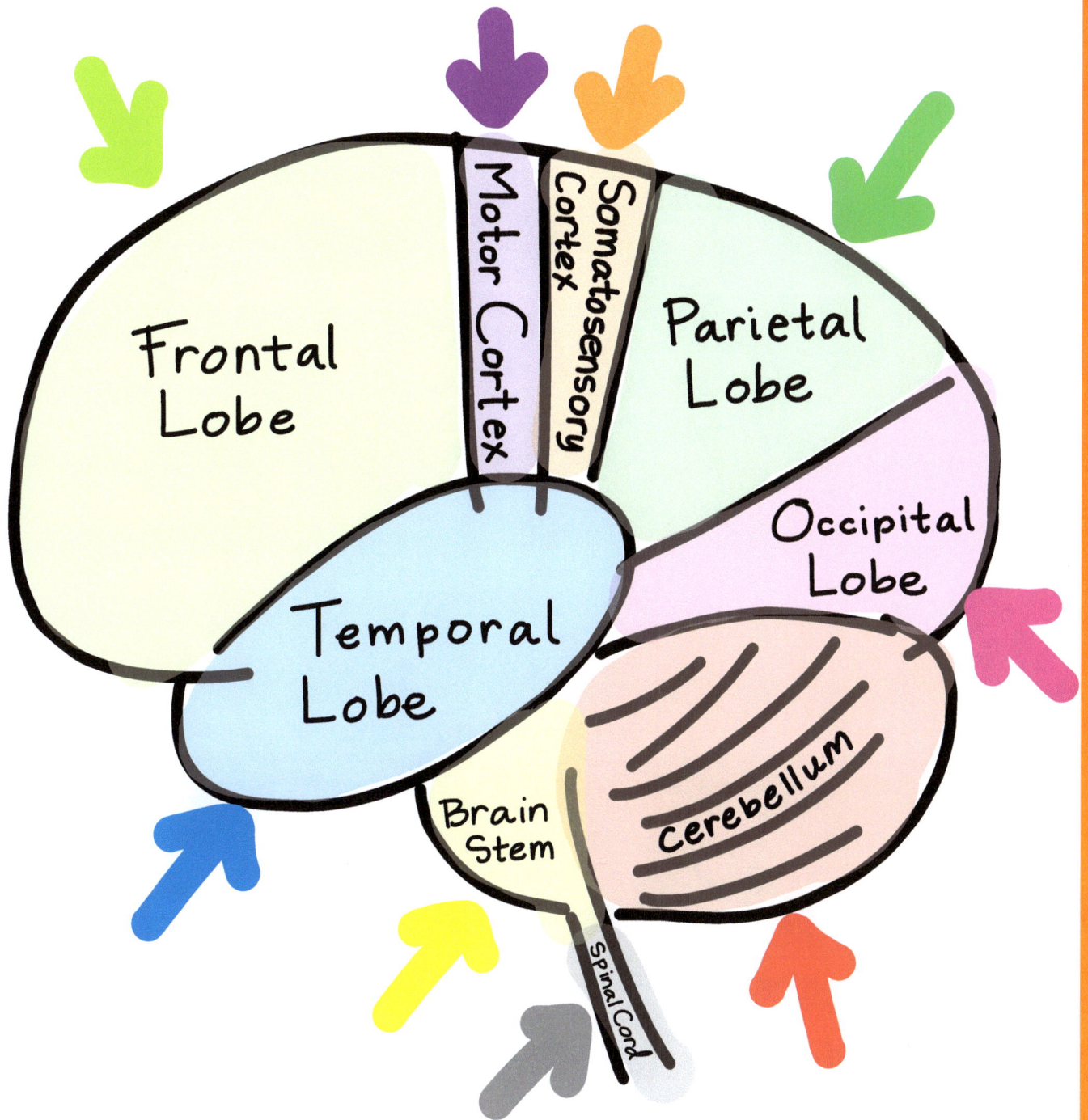

What do you think each part of the brain does?

Are YOU ready to

learn about the

the

BRAIN

???

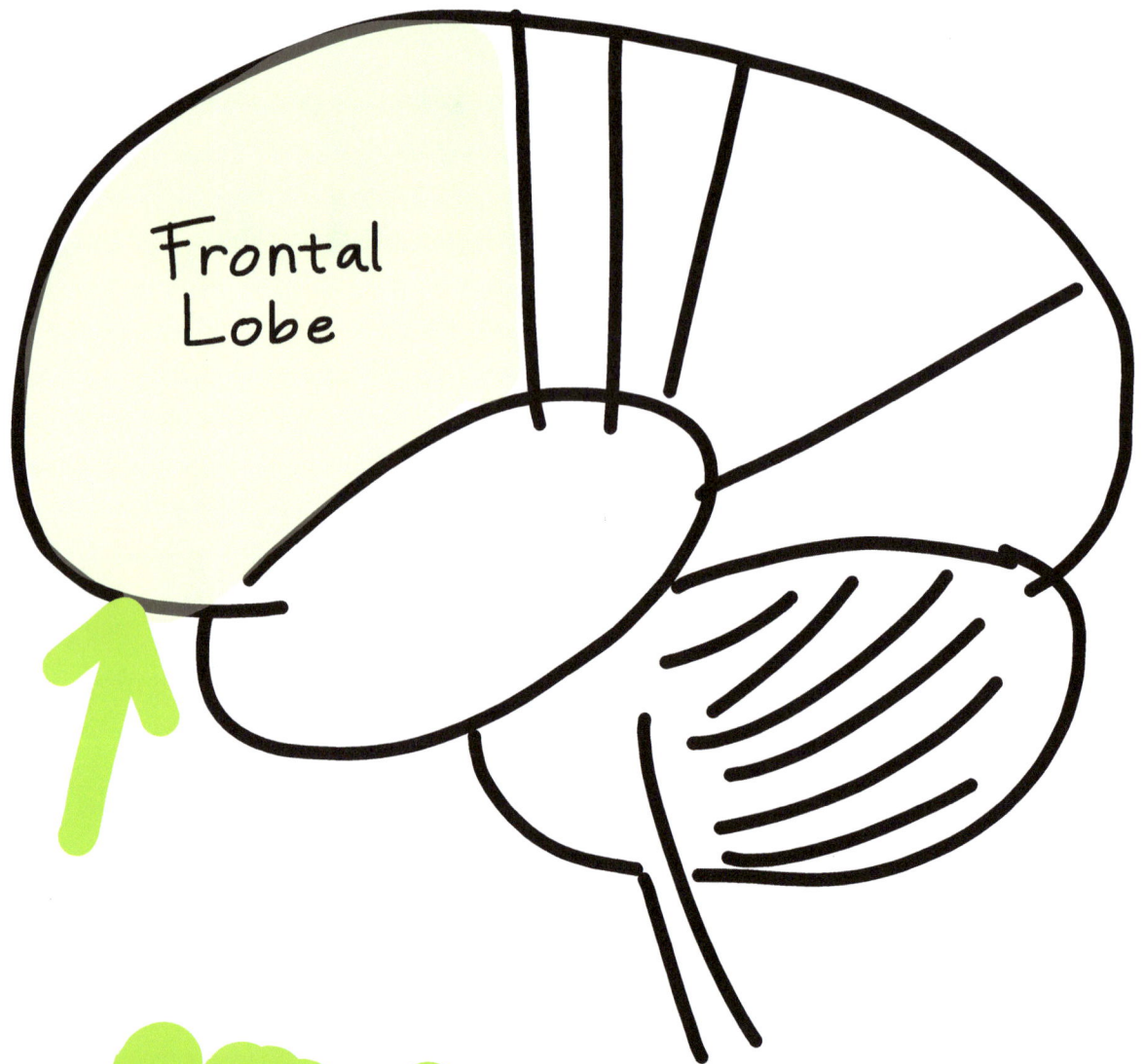

Frontal Lobe

Frontal Lobe

Sound it Out

1. **FRUN**
2. **TUL**
1. **LOB**

Touch your forehead.

The **Frontal Lobe** is the part of your **brain** at your forehead.

The **Frontal Lobe**
<u>**makes you speak and move**</u>.

Say "Hello. How are you today?"
Move your arms like a bird.

That is your **Frontal Lobe** working!

The **Frontal Lobe** also helps you solve problems like this one:
100 + 100 = ?

Can you think of another
problem that your
Frontal Lobe helps you solve?

EXCELLENT WORK!

Remember this:
Frontal = Forehead

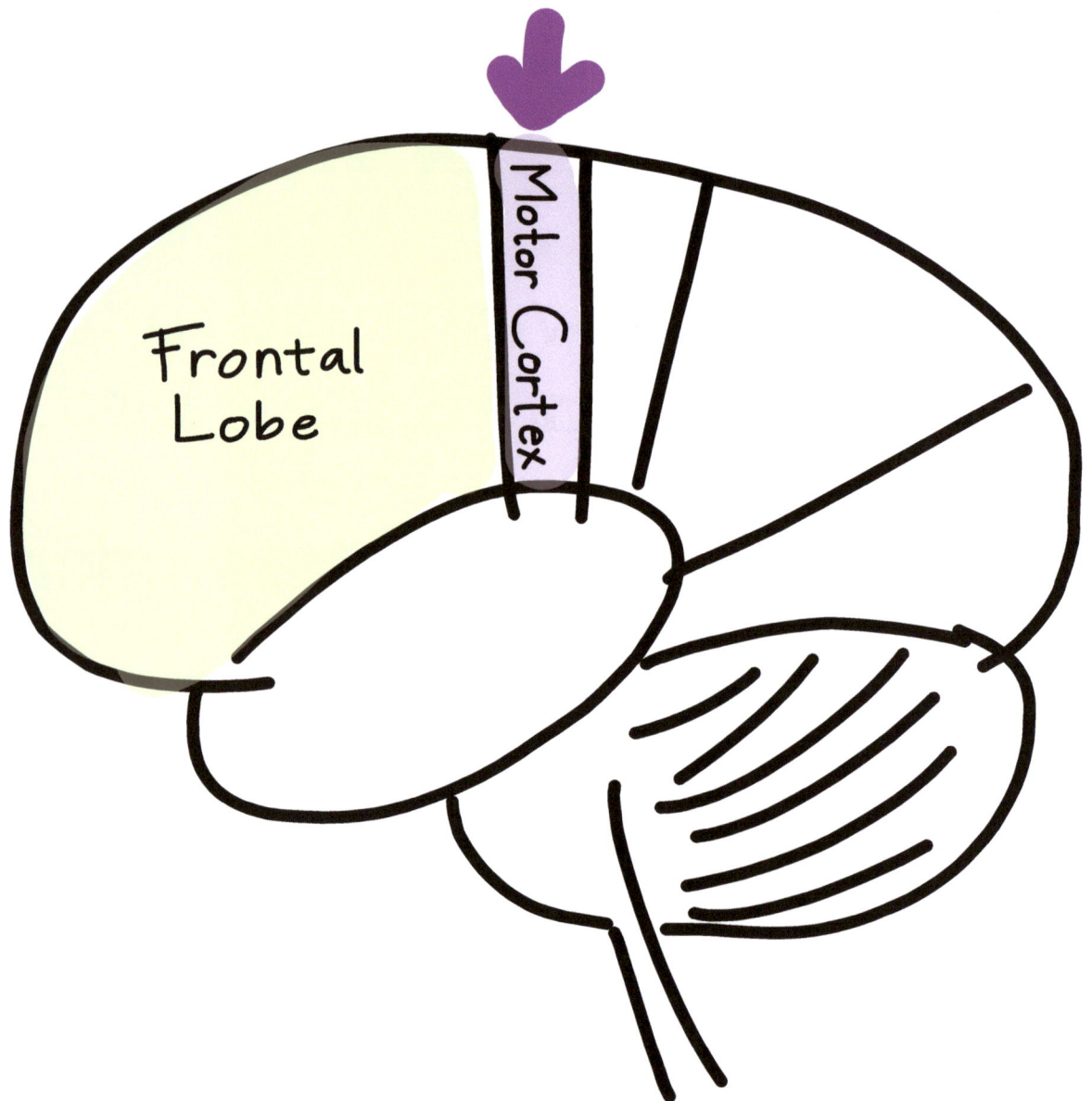

Frontal Lobe

Motor Cortex

Motor Cortex

Sound it Out
1. MO
2. TR

Sound it Out
1. KOR
2. TEX

The Motor Cortex controls voluntary movement.

Voluntary movements are any movements that you __think about__ doing, *for example*:

Touch your nose.

Jump up and down.

Reach for the sky.

That is your Motor Cortex working!

What other movements do you think your Motor Cortex controls?

Remember this:
Motor = Move

Frontal Lobe

Motor Cortex

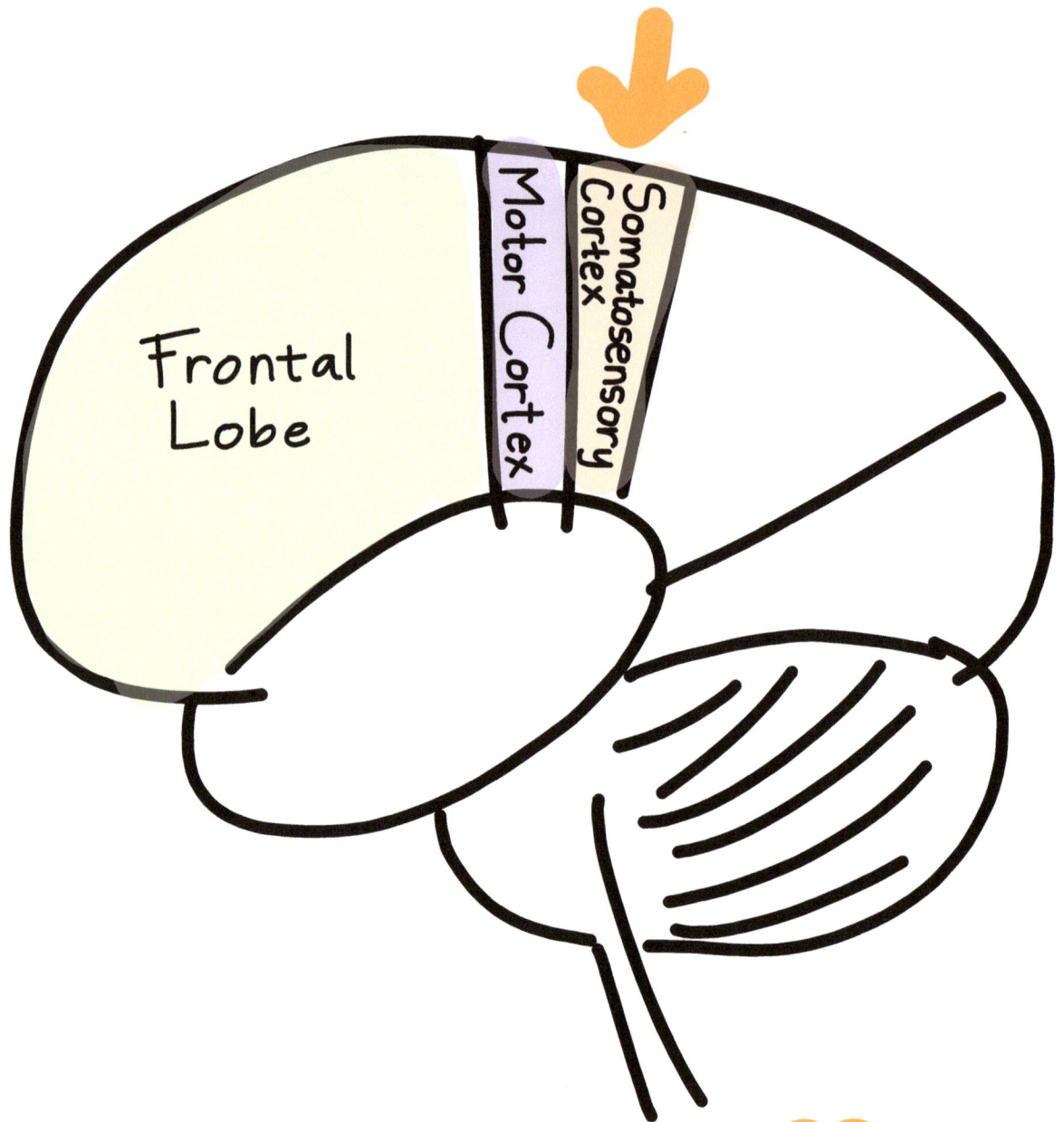

Somatosensory Cortex

Somatosensory Cortex

The **Somatosensory Cortex** <u>receives signals</u> sent from all over your body.

Signals can be sent from your eyes, ears, mouth and nose!

Do you smell flowers**?**
Do you see the bright blue sky**?**
Do you taste yummy food**?**

The signals from all of these are sent to the Somatosensory Cortex of your brain!

What other signals do you think your Somatosensory Cortex receives?

Remember this:
Somatosensory = Signal

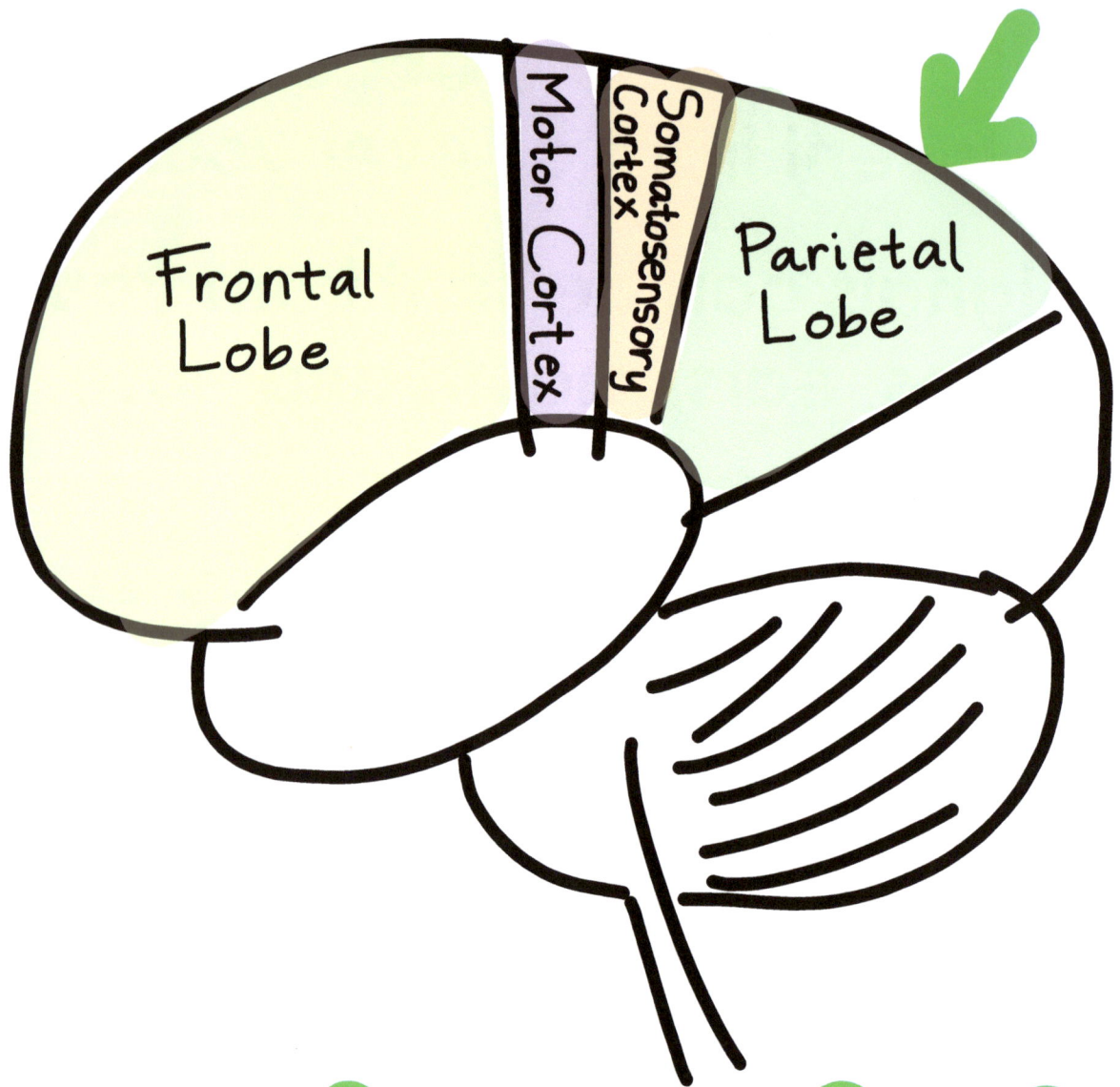

Frontal Lobe

Motor Cortex

Somatosensory Cortex

Parietal Lobe

Parietal Lobe

Sound it Out

1. PA
2. RI
3. EH
4. TUL

Sound it Out

1. LOB

The **Parietal Lobe** <u>senses</u> things like touch, temperature and pain.

Do you know what paper feels like**?**
Can you feel the heat from the sun**?**

You are able to <u>sense</u> these things because the **Parietal Lobe** of your **brain** is hard at work.

What other sensations do you think your Parietal Lobe processes?

Remember this:
Parietal = Sensation

Review the sections of the brain you have learned so far.

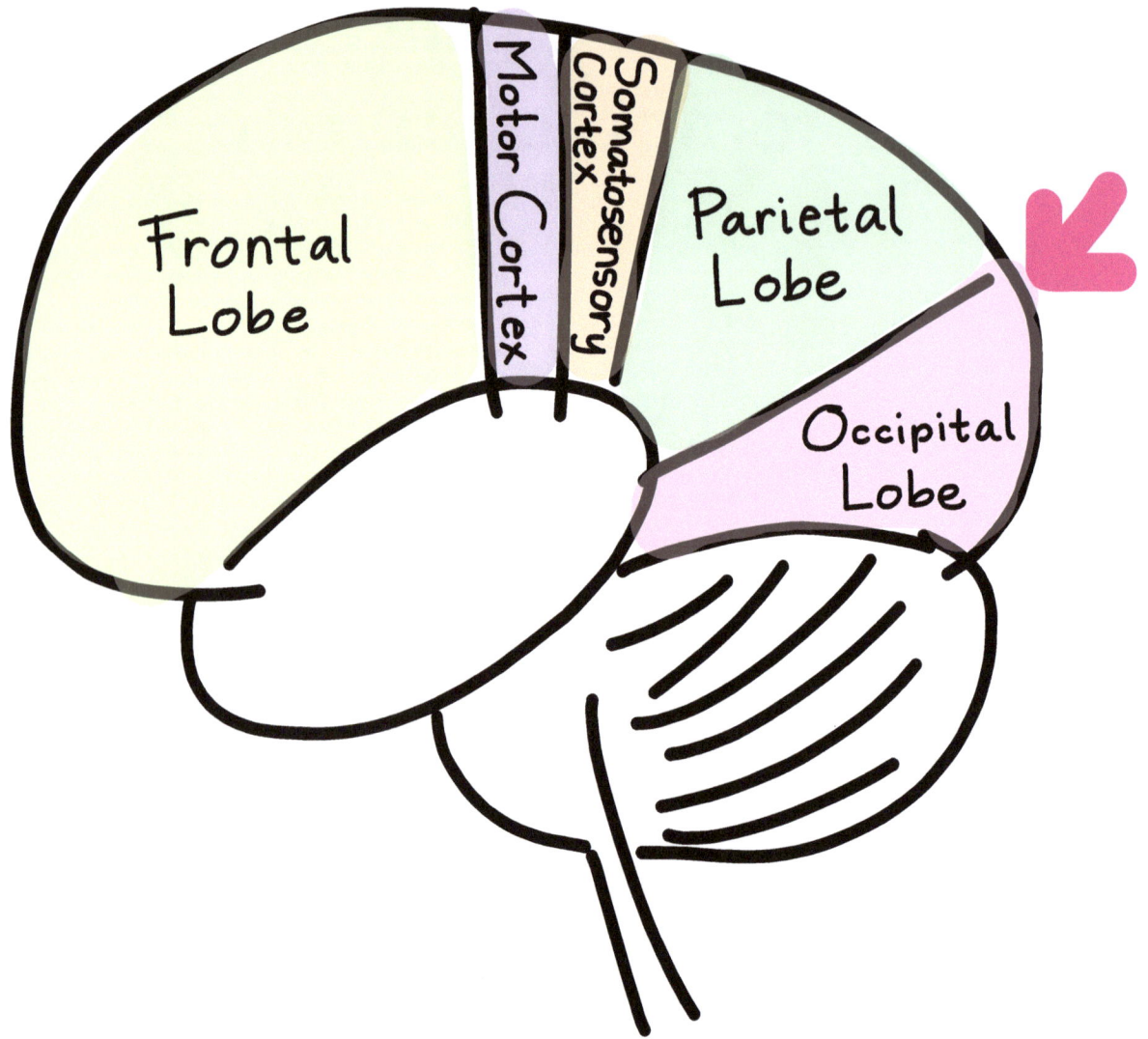

Frontal Lobe

Motor Cortex

Somatosensory Cortex

Parietal Lobe

Occipital Lobe

Occipital Lobe

Sound it Out

1. OX
2. IP
3. EH
4. TUL

Sound it Out

1. LOB

The **Occipital Lobe** <u>processes</u> the signals from your eyes so you <u>understand</u> what you are seeing.

Do you see the difference between Pink and Blue?

Do you know what a dog looks like?

You can understand what you are seeing because of the **Occipital Lobe** of your **brain**!

What other signals do you think your **Occipital Lobe** helps you understand?

Remember this:
Occipital = Understand

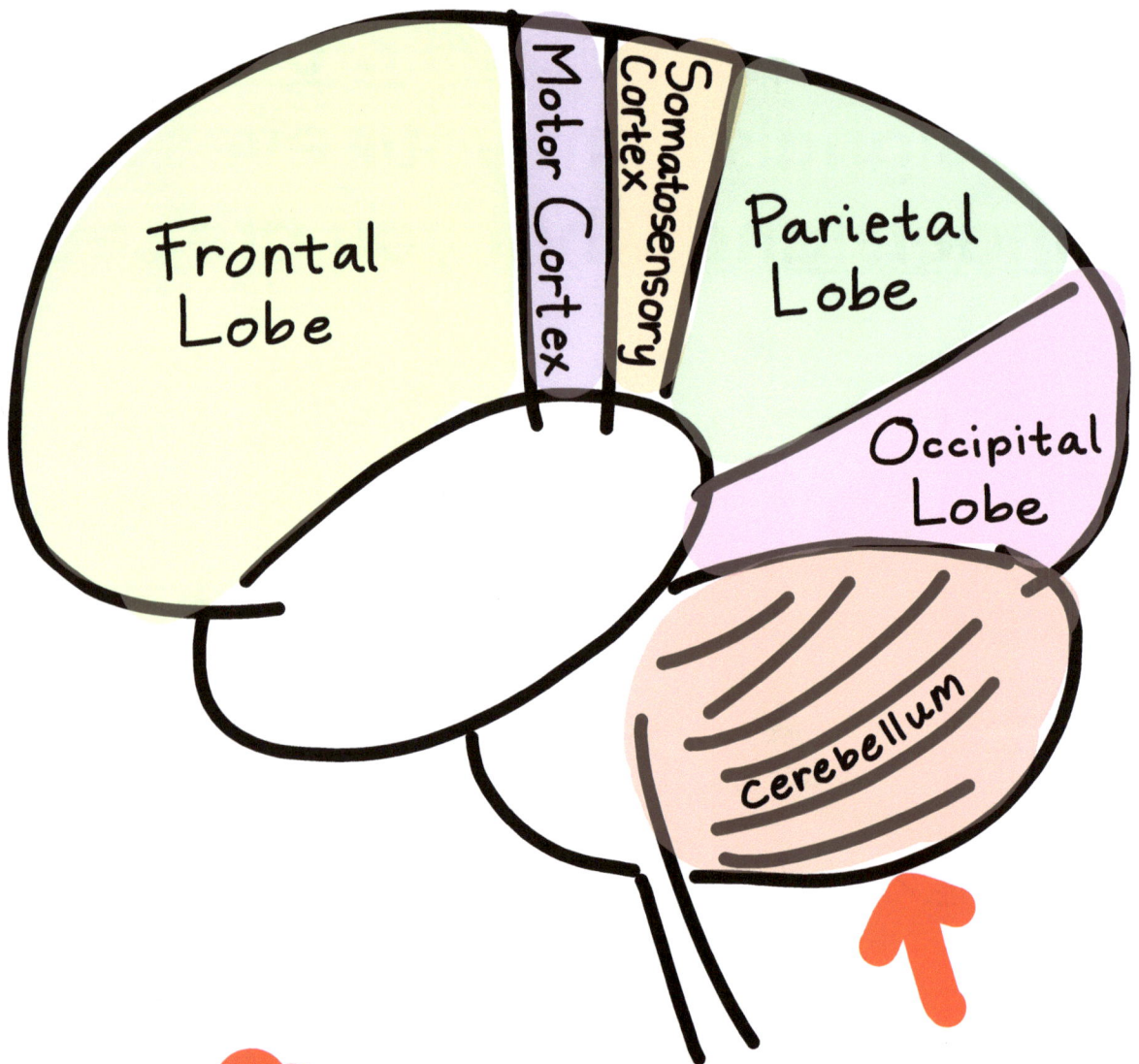

Frontal Lobe

Motor Cortex

Somatosensory Cortex

Parietal Lobe

Occipital Lobe

Cerebellum

Cerebellum

Sound it Out
1. SER
2. EH
3. BEL
4. UM

Touch the back of your head by your neck.

This is where your Cerebellum is located.

The **Cerebellum** <u>controls balance</u>, movement and coordination (how your muscles all work together).

Do you know how to ride a bike**?**

Have you ever seen a surfer on a big wave**?**

The **Cerebellum** controls balance so you can ride a bike and surf!

What other activities do you think your Cerebellum controls?

Remember this:
Cerebellum = Balance

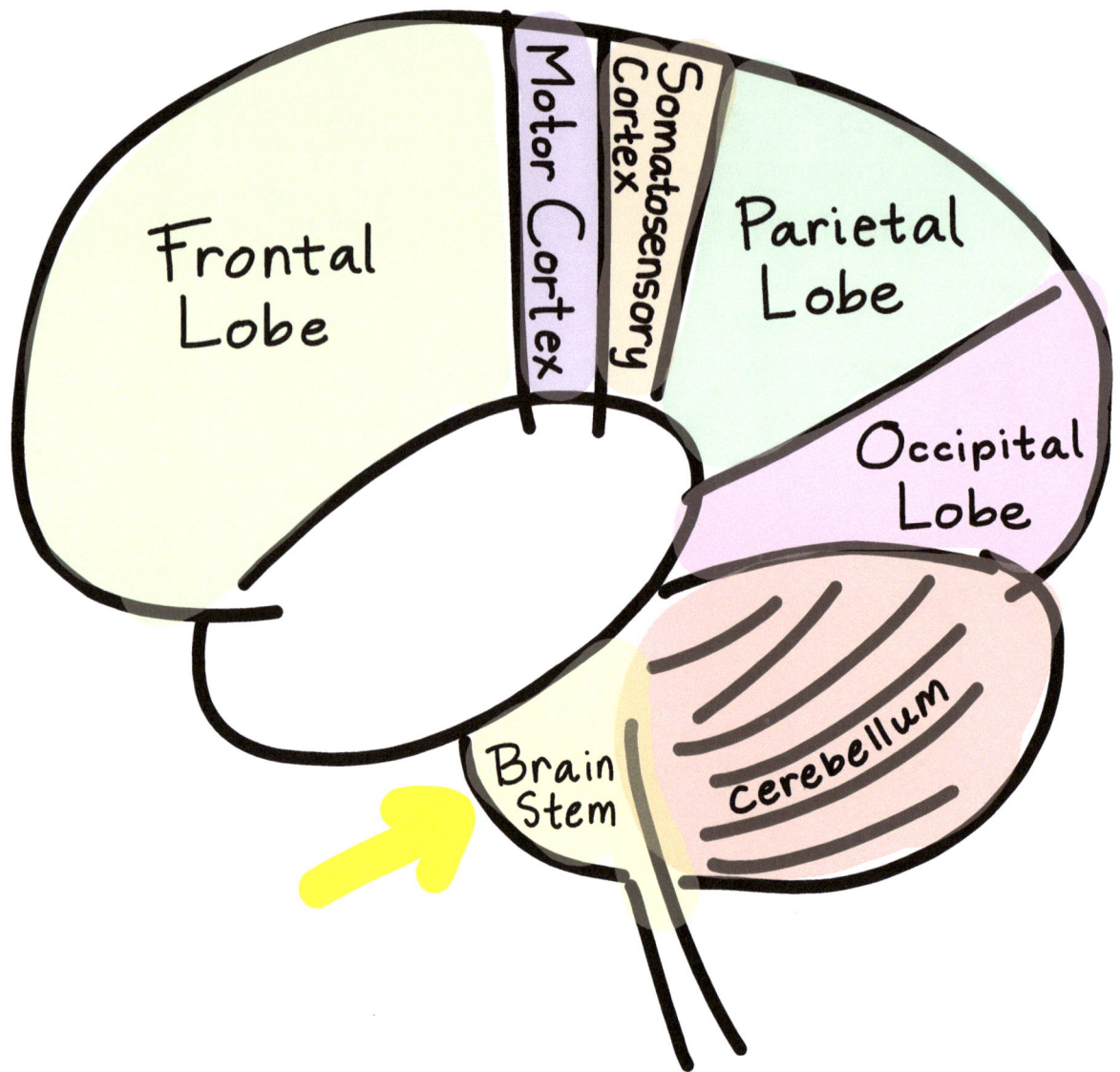

Frontal Lobe

Motor Cortex

Somatosensory Cortex

Parietal Lobe

Occipital Lobe

Brain Stem

Cerebellum

Brain Stem

The **Brain Stem** connects the **brain** to the **spinal cord**.

The **Brain Stem** <u>controls heartbeat</u>, breathing, digestion and waking you up in the morning!

Do you have to think about breathing or making your heart beat**?**

No!

Your Brain Stem controls that action for you.

Remember this:
Brain Stem = Heartbeat

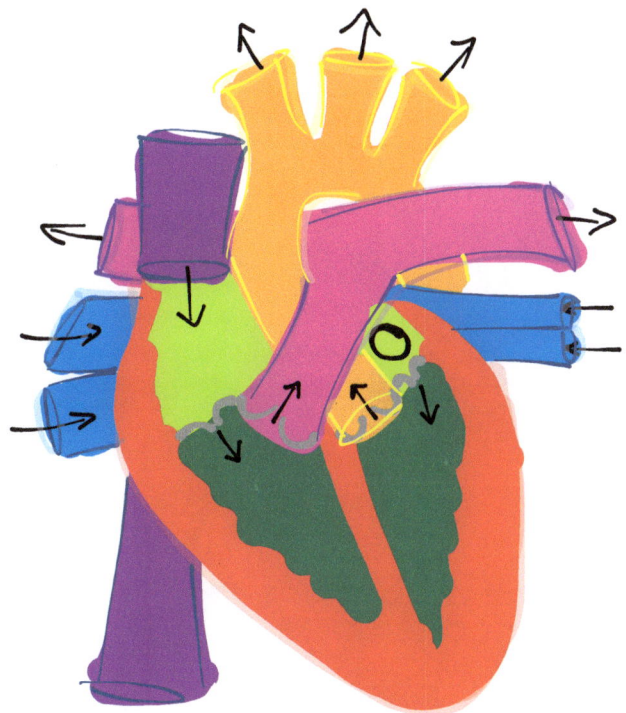

THE HEART

The arrows show the direction of blood flow in and out of the heart. Learn more about the heart in *Cardiology, book 7 of the Super Smart Science Series.*™

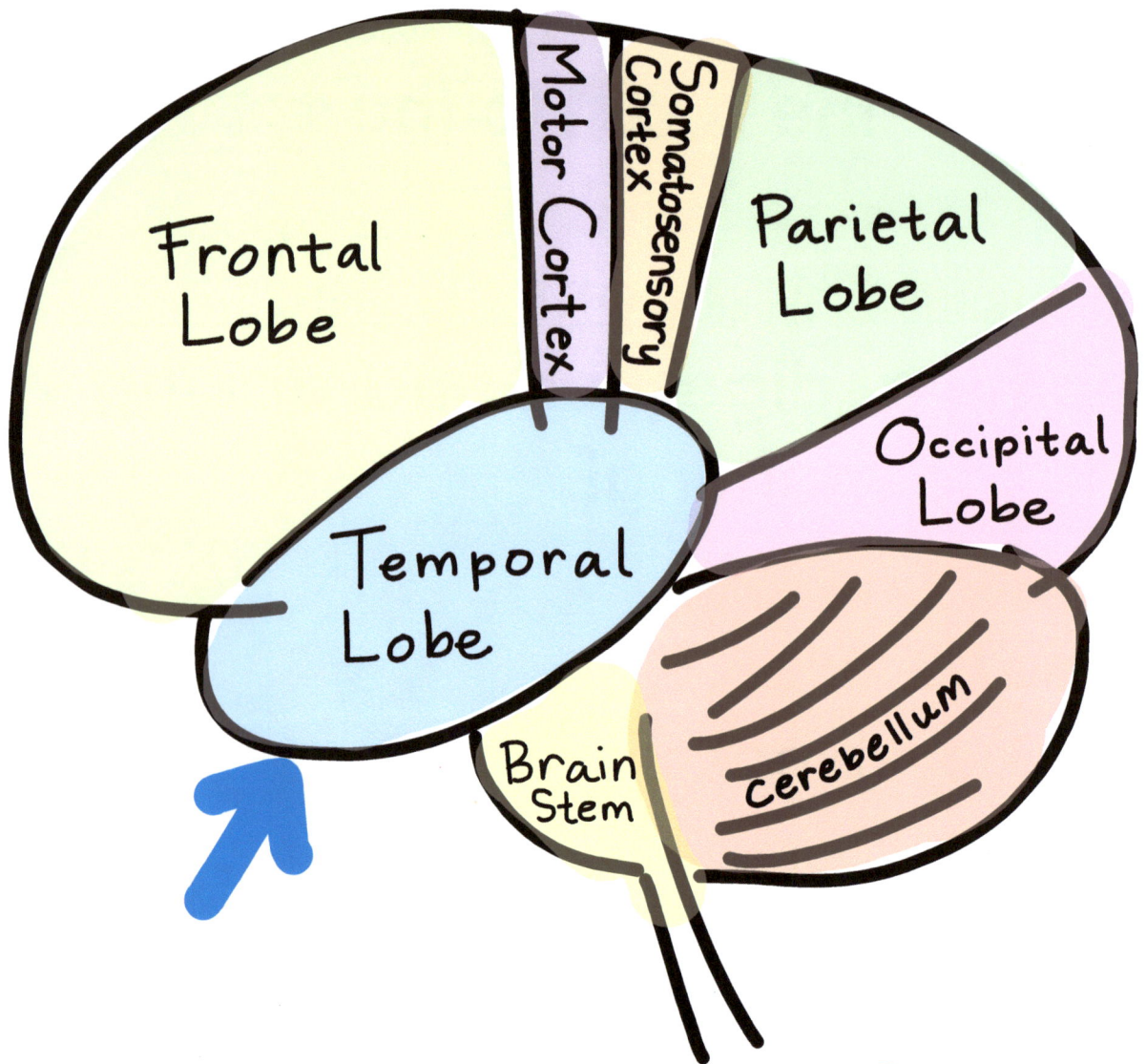

Frontal Lobe

Motor Cortex

Somatosensory Cortex

Parietal Lobe

Occipital Lobe

Temporal Lobe

Brain Stem

Cerebellum

Temporal Lobe

The **Temporal Lobe** <u>controls hearing.</u> It receives sounds and speech and comprehends what you are hearing.

Can you hear a bird chirp?

Do you recognize that sound as being a bird chirp?

You can hear and understand what you are hearing because of the Temporal Lobe of your brain.

What other things does your Temporal Lobe allow you to hear?

Remember this:
Temporal = Ears

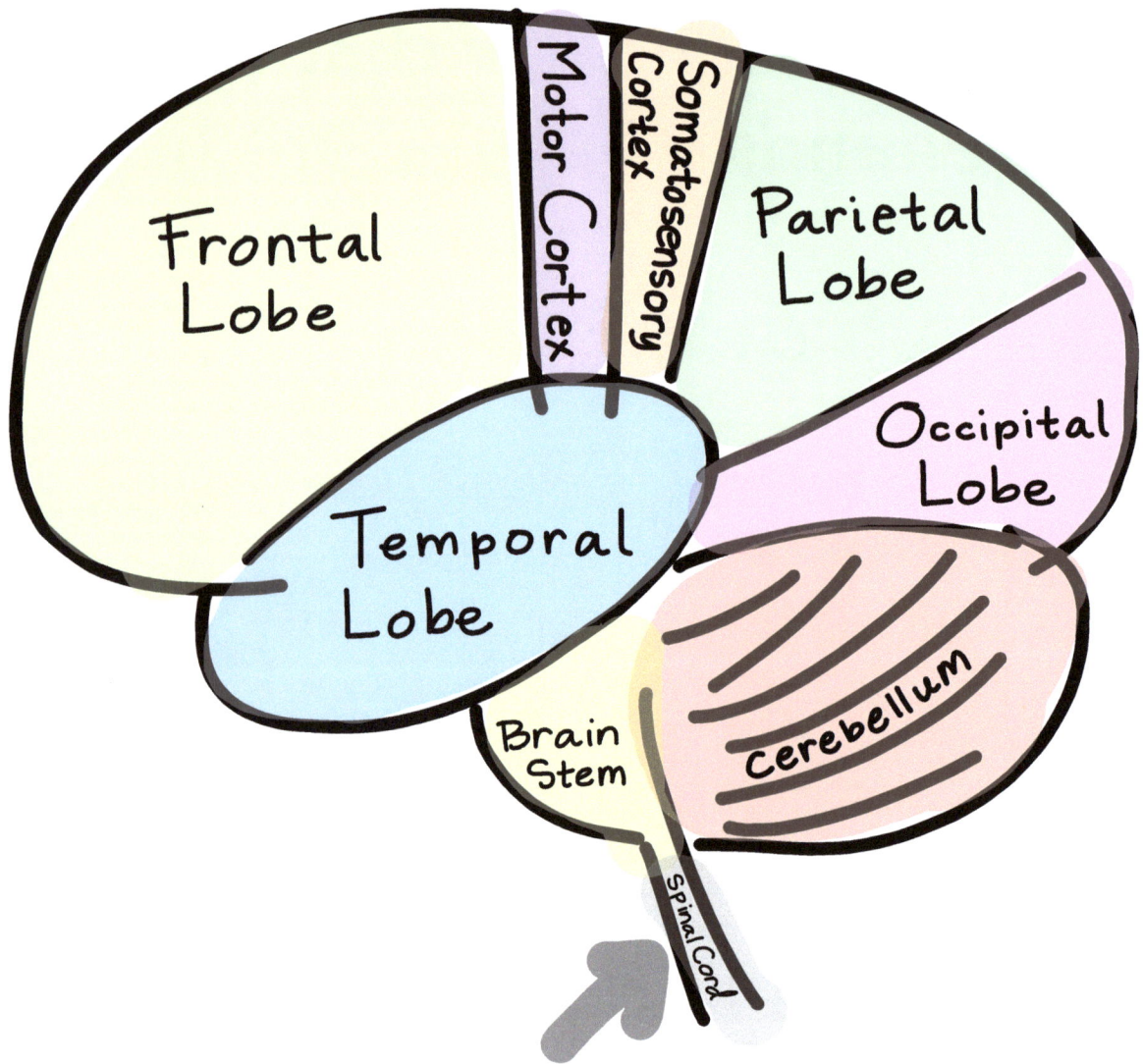

Frontal Lobe

Motor Cortex

Somatosensory Cortex

Parietal Lobe

Occipital Lobe

Temporal Lobe

Brain Stem

Cerebellum

Spinal Cord

Spinal Cord

Sound it Out

1. SPI
2. NUL

Sound it Out

1. KORD

The **Central Nervous System** is the
BRAIN + SPINAL CORD

The spinal cord starts at the base of the brain next to the brain stem and continues all the way down your back.

The spinal cord is the pathway for messages between the brain and body.

The spinal cord is protected by bumpy bones called vertebrae.

Vertebrae

Sound it Out

1. VER
2. TE
3. BRAY

You have:

8 cervical vertebrae,
Sound it out: SER-VI-KUL

12 thoracic vertebrae,
Sound it out: THOR-AS-IK

5 lumbar vertebrae,
Sound it out: LUM-BAR

5 sacral vertebrae
Sound it out: SA-KRUL

and
1 coccygeal vertebra!
Sound it out: KOX-E-JE-UL

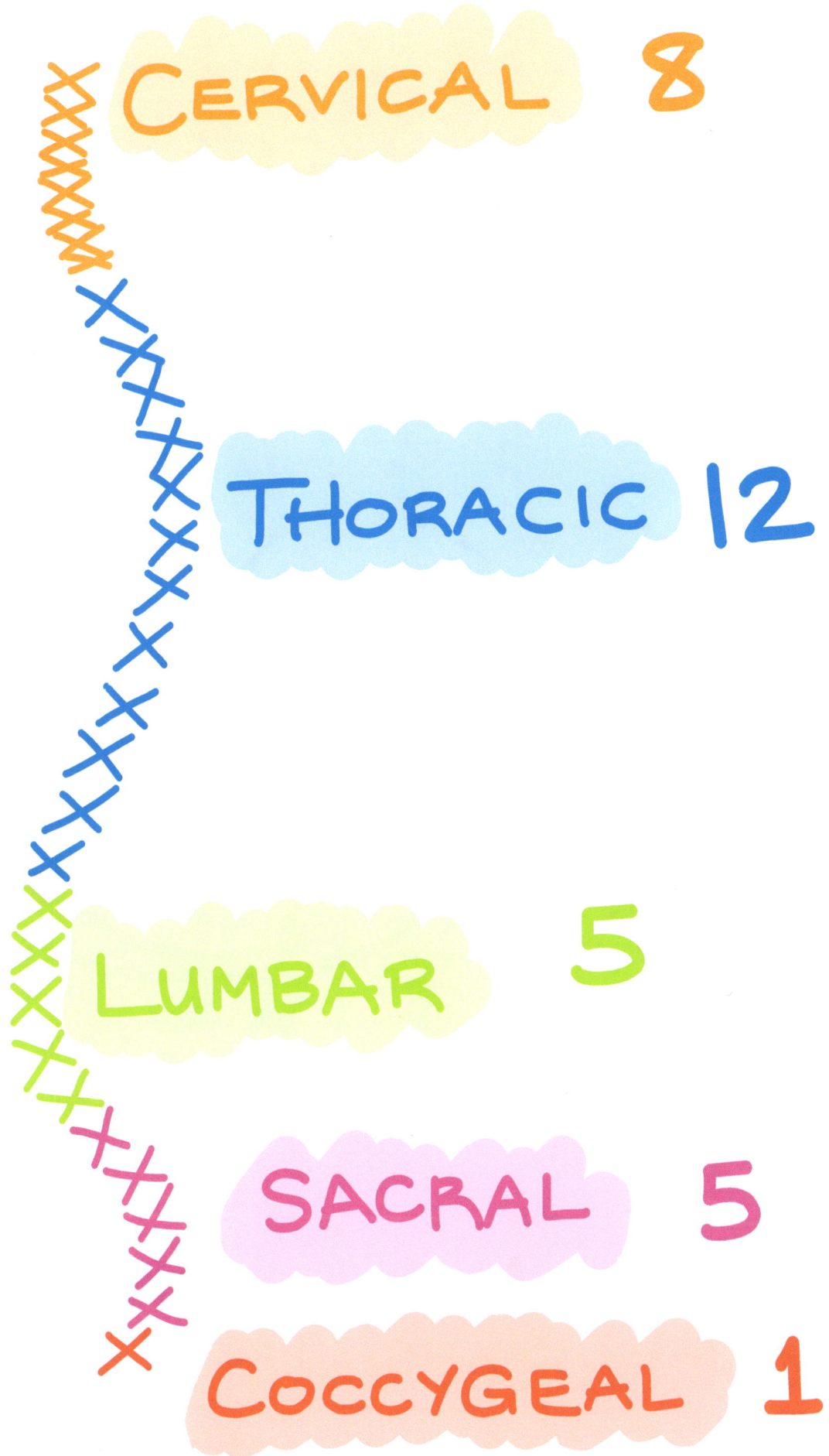

CERVICAL 8

THORACIC 12

LUMBAR 5

SACRAL 5

COCCYGEAL 1

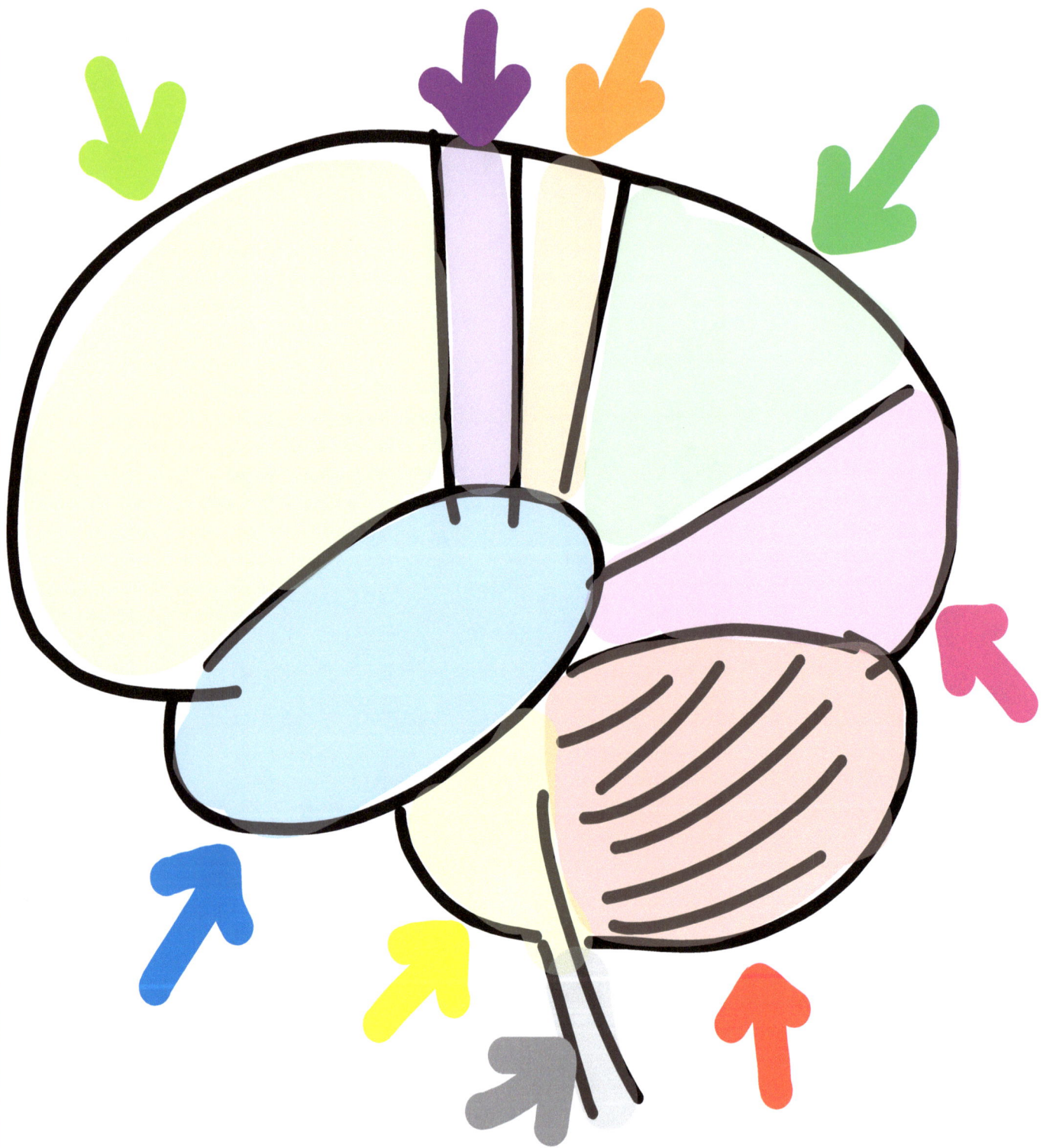

**Match each
colored section in the picture
to the correct name.**

Frontal Lobe
Motor Cortex
Somatosensory Cortex
Parietal Lobe
Occipital Lobe
Cerebellum
Brain Stem
Temporal Lobe
Spinal Cord

Do you remember the function of each part of the brain?

GOOD JOB!

Who wants to learn more?!

Remember, the
Central Nervous System is the
BRAIN + SPINAL CORD.

What is the spinal cord made of?

Bundles of Neurons!

Neuron

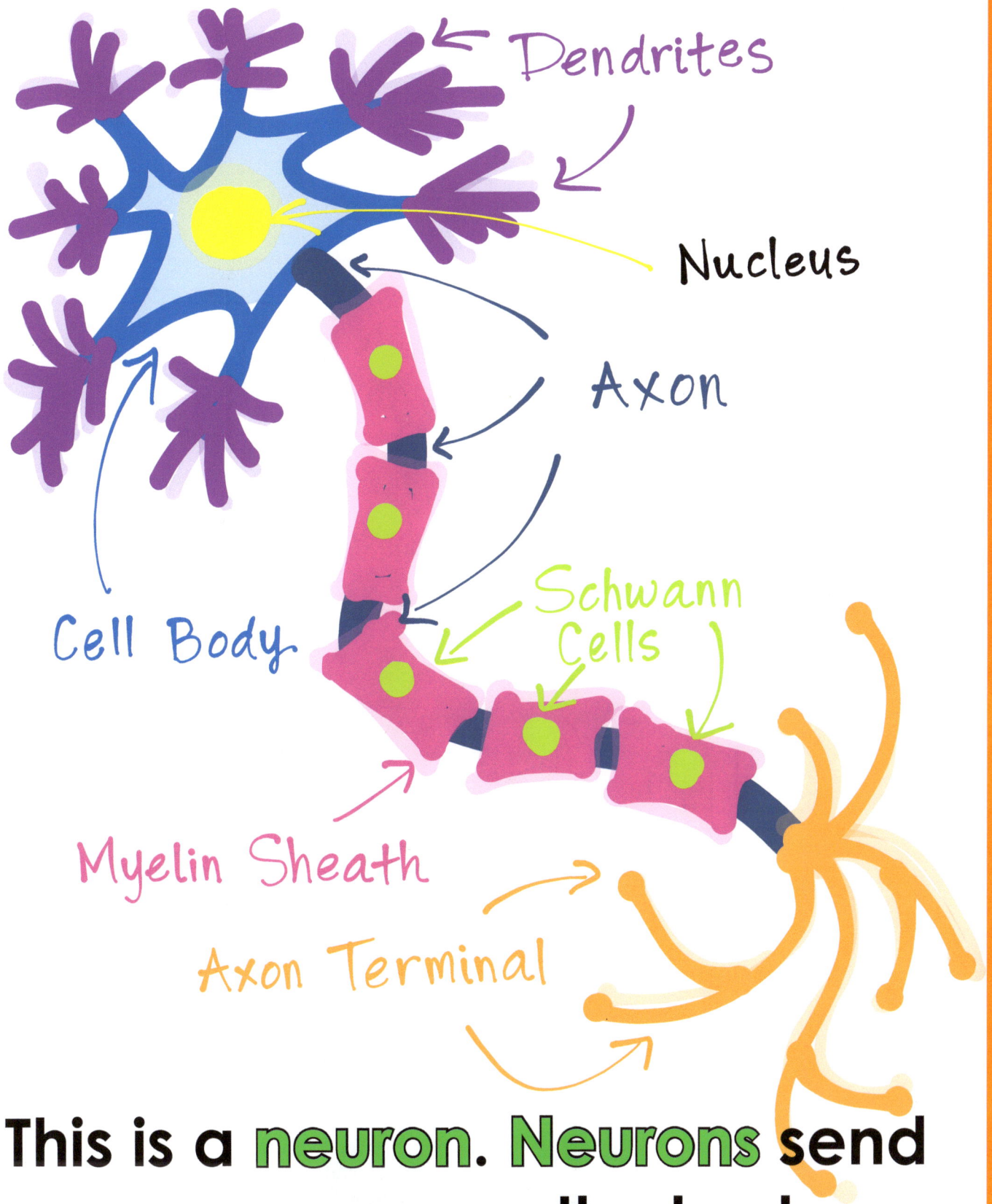

Dendrites

Nucleus

Axon

Schwann Cells

Cell Body

Myelin Sheath

Axon Terminal

This is a neuron. Neurons send messages across the body. What is the function of each part of the neuron? Let's see!

Nucleus

Nucleus

The **nucleus** contains DNA.

The **nucleus** is located inside something... inside what?

Turn the page and find out!

(The terms "NUCLEUS" and "DNA" also appear in Cellular Biology: Organelles, Structure, Function, Book 1 of the Super Smart Science Series™)

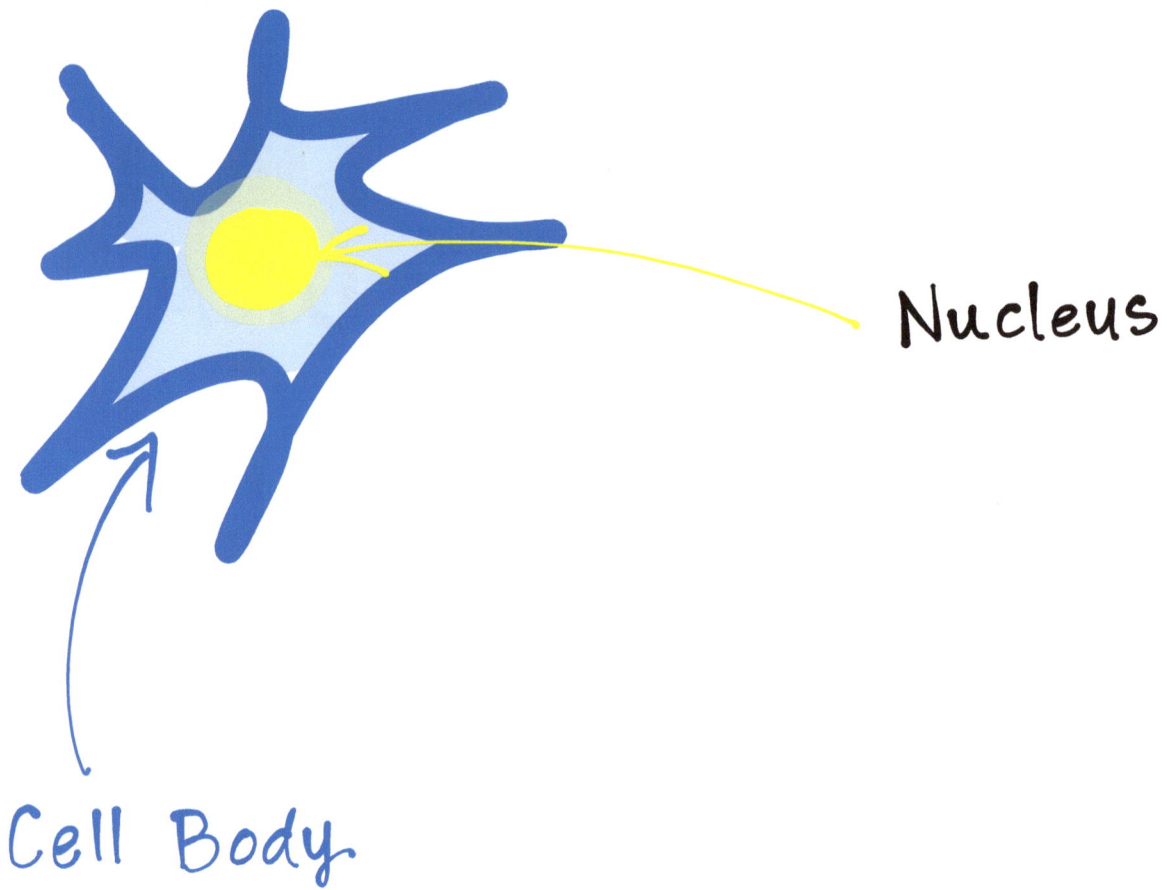

Nucleus

Cell Body

Cell Body

The **nucleus**
is located inside
the **cell body**!

The **cell body** receives
a message from a part
of the **neuron**, then
sends it to another part.

Where is this message
coming from?
And where does it go?

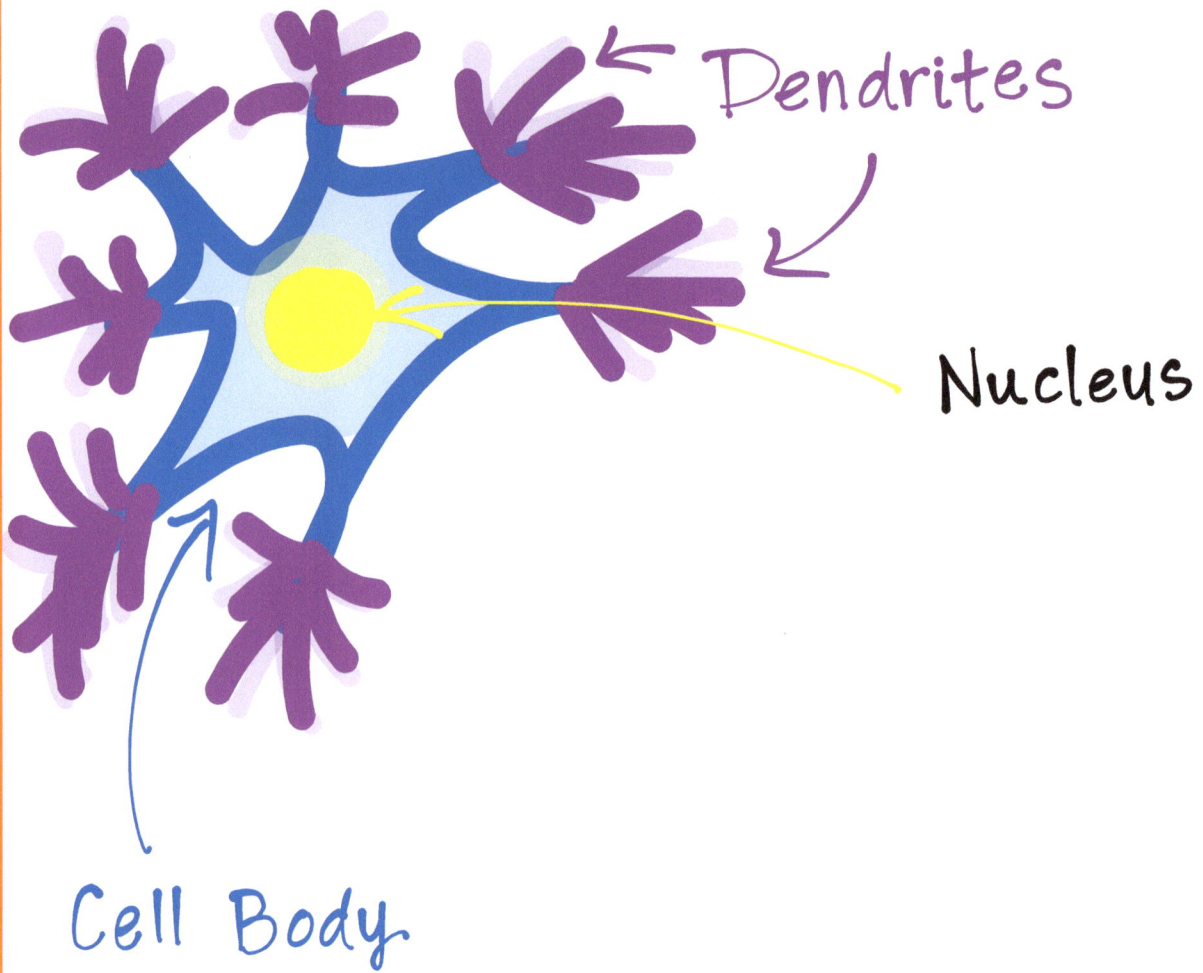

Dendrites

Nucleus

Cell Body

Dendrites

The message comes from the dendrites.

The dendrites send the message to the cell body.

dendrites

cell body

_____

Where does the message go from there?

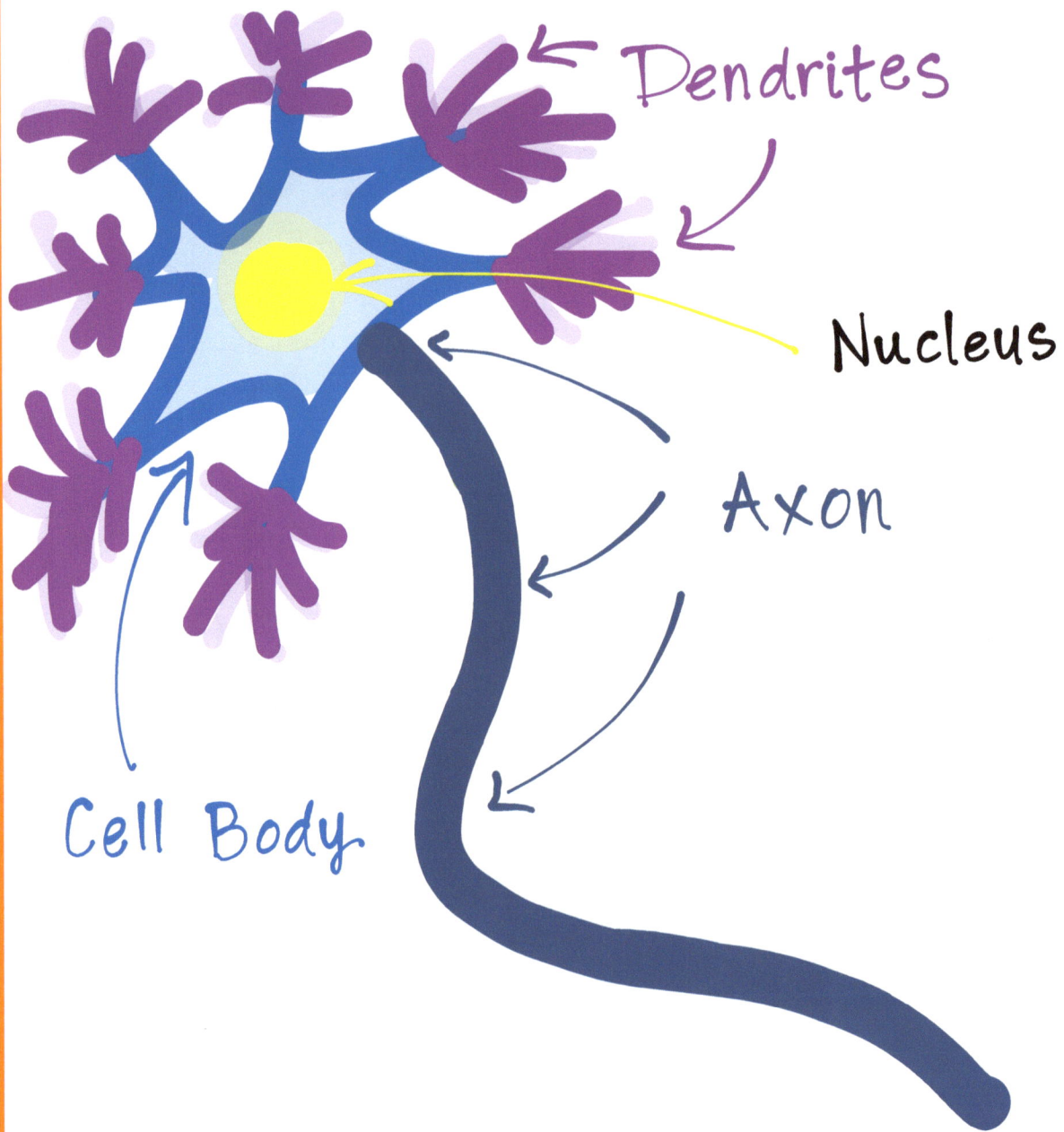

Dendrites

Nucleus

Axon

Cell Body

Axon

The message leaves the **cell body** and goes down the **axon**.

Review the path of the message so far:

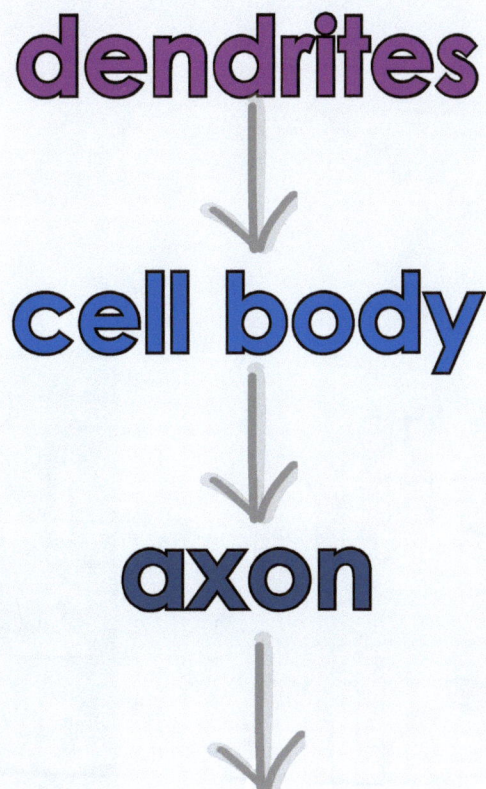

dendrites

↓

cell body

↓

axon

↓

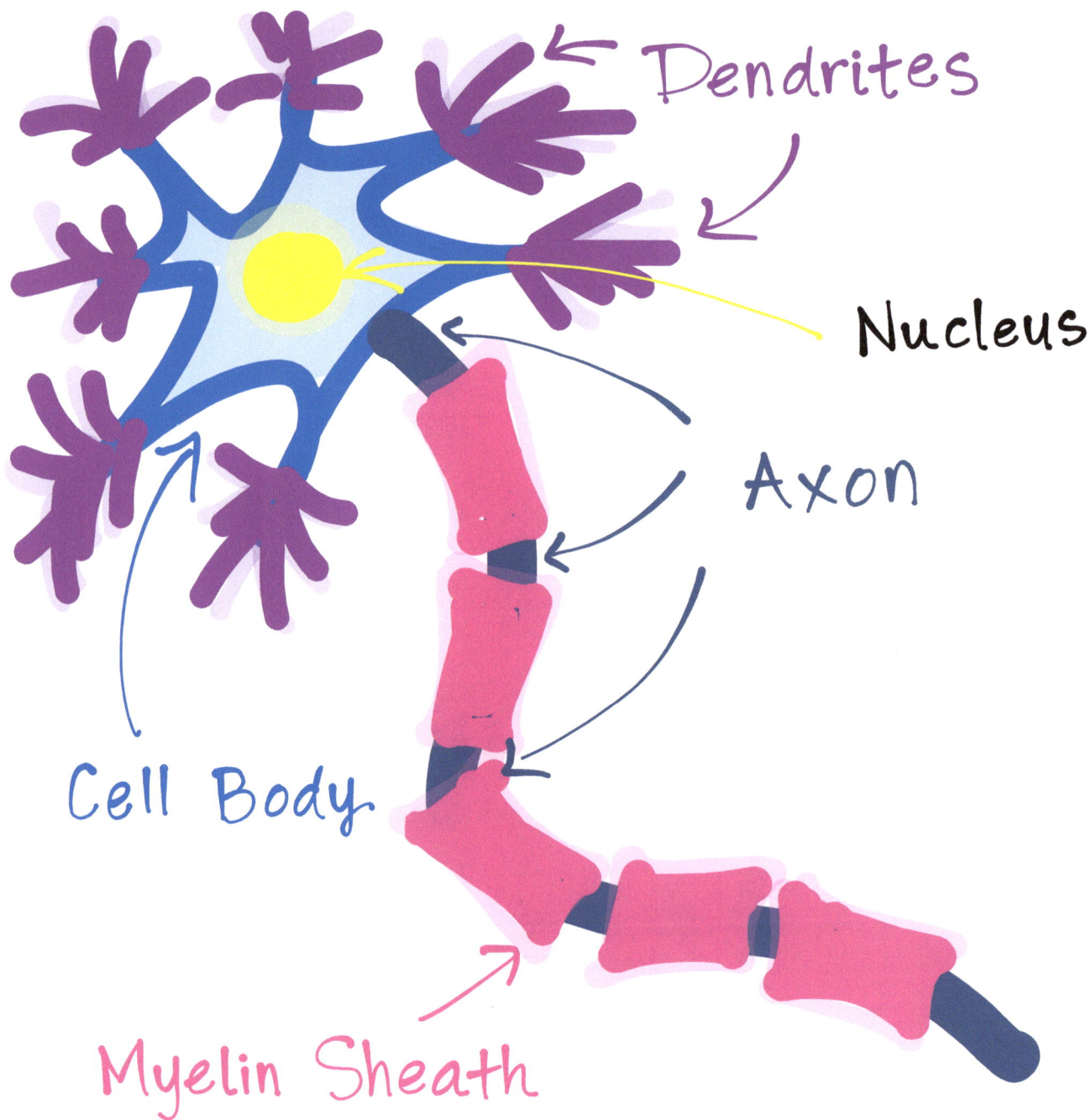

Dendrites

Nucleus

Axon

Cell Body

Myelin Sheath

Myelin Sheath

The **myelin sheath** wraps around the **axon** to help the message send quicker.

The **myelin sheath** does not cover the entire **axon**.

Do you see the **axon** between the **myelin sheaths**?

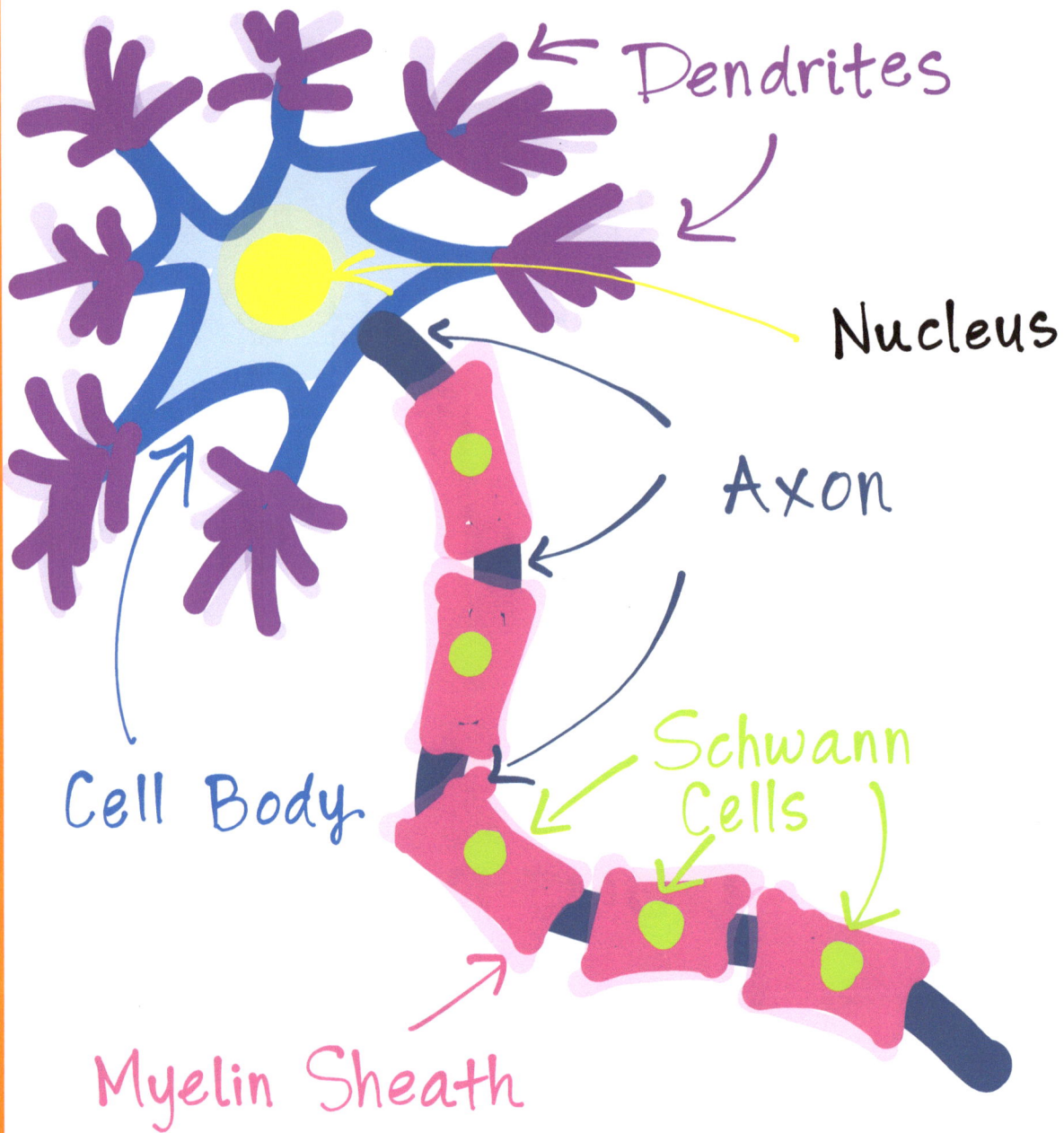

Dendrites

Nucleus

Axon

Schwann Cells

Cell Body

Myelin Sheath

Schwann Cells

Sound it Out

1. SHWAN
2. SELLS

Schwann cells <u>make</u> the **myelin sheaths** that wrap around the **axon**.

Do you remember what the **myelin sheath** does?

What is the path of the message from the beginning?

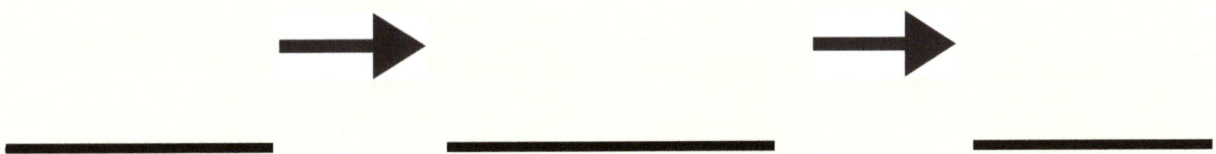

_____ → _____ → _____

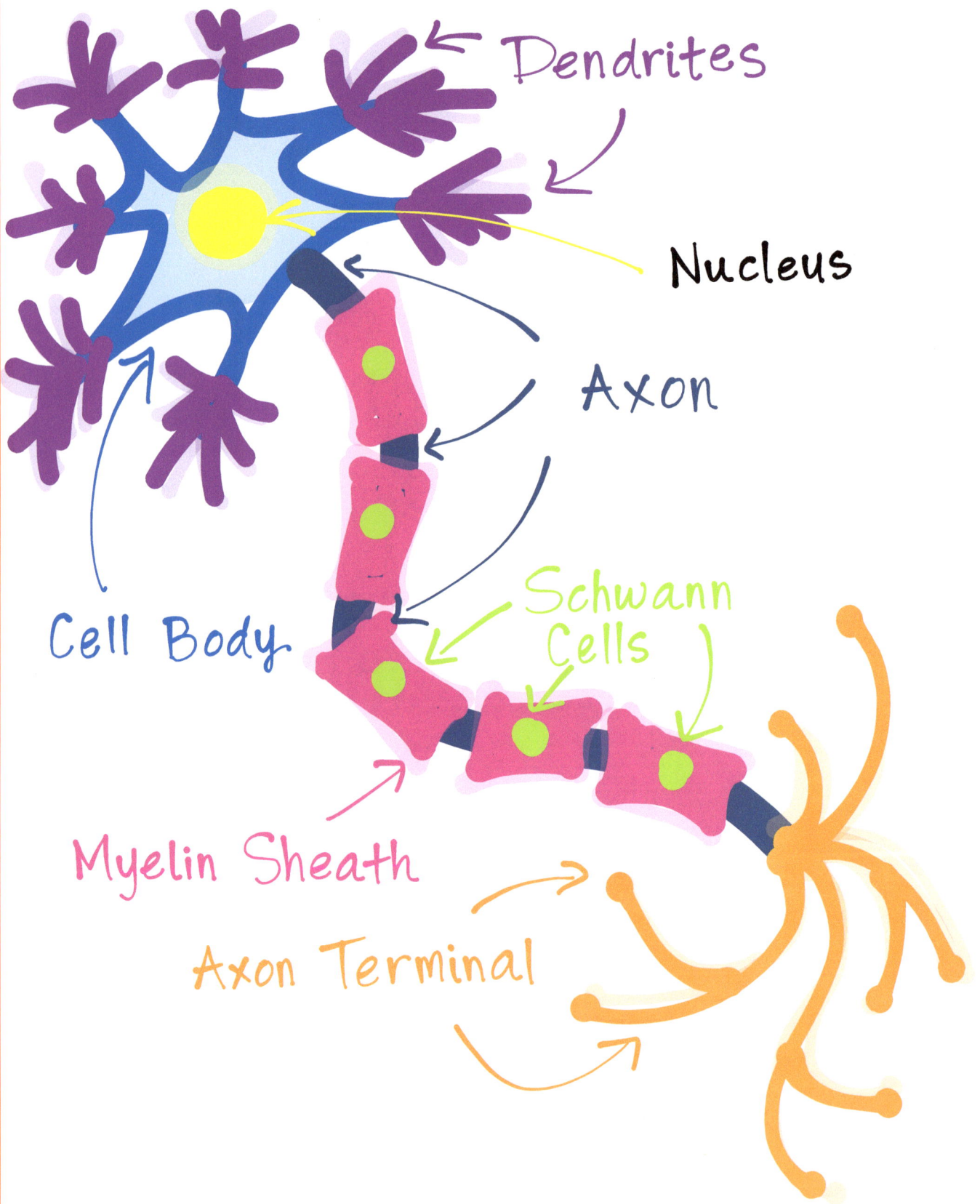

Dendrites

Nucleus

Axon

Cell Body

Schwann Cells

Myelin Sheath

Axon Terminal

Axon Terminal

Sound it Out
1. AX
2. ON

1. TUR
2. MIN
3. UL

The axon terminal receives the message from the axon.

The axon terminal sends the message to the next neuron's dendrites.

Dendrites
↓
Cell Body
↓
Axon
↓
Axon Terminal
↓
Dendrites
↓

Neurons line up and send the message from one to another.

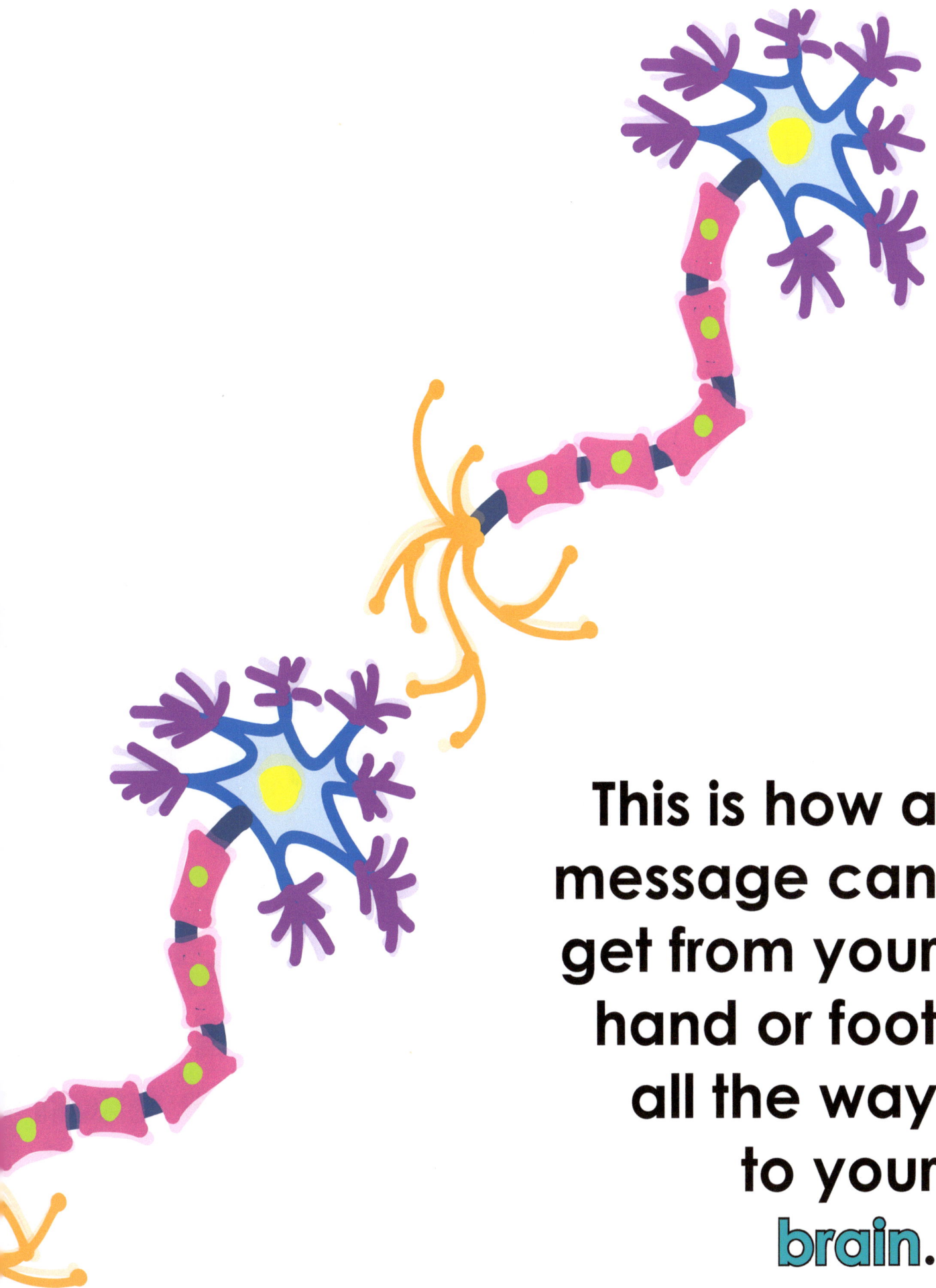

This is how a message can get from your hand or foot all the way to your brain.

"Remember this" REVIEW

Frontal=**?**
Motor=**?**
Somatosensory=**?**
Parietal=**?**
Occipital=**?**
Cerebellum=**?**
Brain Stem=**?**
Temporal=**?**

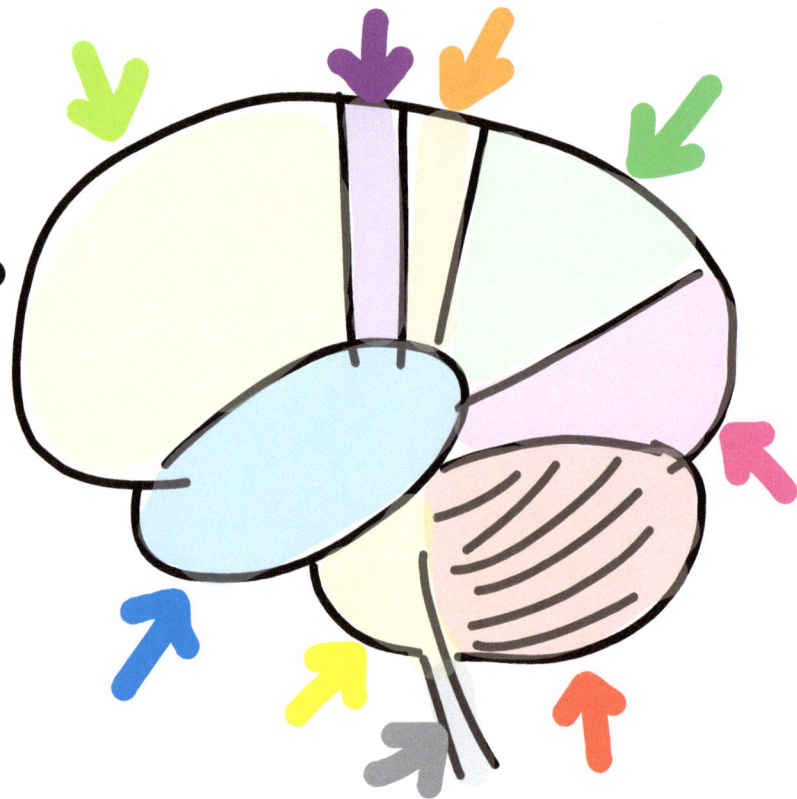

Neurons are the connections between your brain and your body.

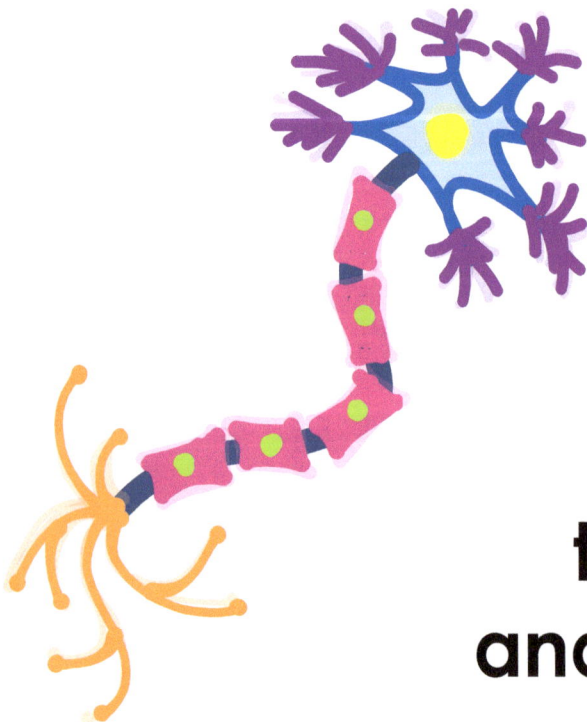

Messages are sent from the dendrites to the cell body, then down the axon to the axon terminal, and to the next neuron.

The **spinal cord** is made of bundles of **neurons**.

It is the pathway for messages between the **brain** and body.

The **spinal cord** is protected by bumpy bones called **vertebrae**.

You are a Neurology expert!

Astronomy:
The Solar System

Ages 0-100

Neptune

Jupiter

Venus

Saturn

The Sun

Mercury

Earth

Mars

Uranus

Book FOUR of the
Super Smart Science Series ™

By: APRIL CHLOE TERRAZAS

Benny Rodriguez

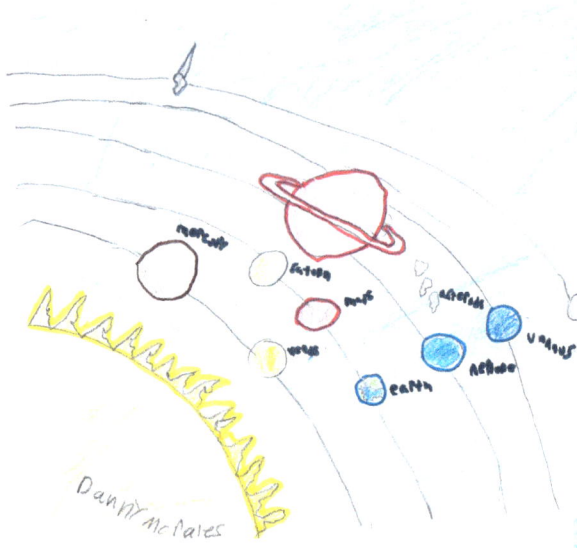

Danny Morales

Bui Ariam Pantoja

Yareli

Astronomy: The Solar System

Written and Illustrated by: APRIL CHLOE TERRAZAS

This book is dedicated to:

My Super Awesome Cousins Jimito & Mike Ruiz

Through y'all, I have truly been able to reach for the stars!

Astronomy: The Solar System. April Chloe Terrazas, BS University of Texas at Austin.
Copyright © 2013 Crazy Brainz, LLC

Visit us on the web! www.Crazy-Brainz.com

Cover design, illustrations and text by: April Chloe Terrazas

Some super cool terms to know:

Orbit

Sound it Out	The curved path of an object
1. OR	in space around a star,
2. BIT	**planet** or moon. Do you see the moon **orbiting** Earth?

Planet

Sound it Out	A large object in space that
1. PLAN	**orbits** around a star
2. IT	(like the **Sun**).
	Do you see **planet** Earth?

Crater

A large bowl-shaped area on the surface of a planet or moon. It is usually caused by an explosion or impact of a meteorite. Do you see the craters on the moon?

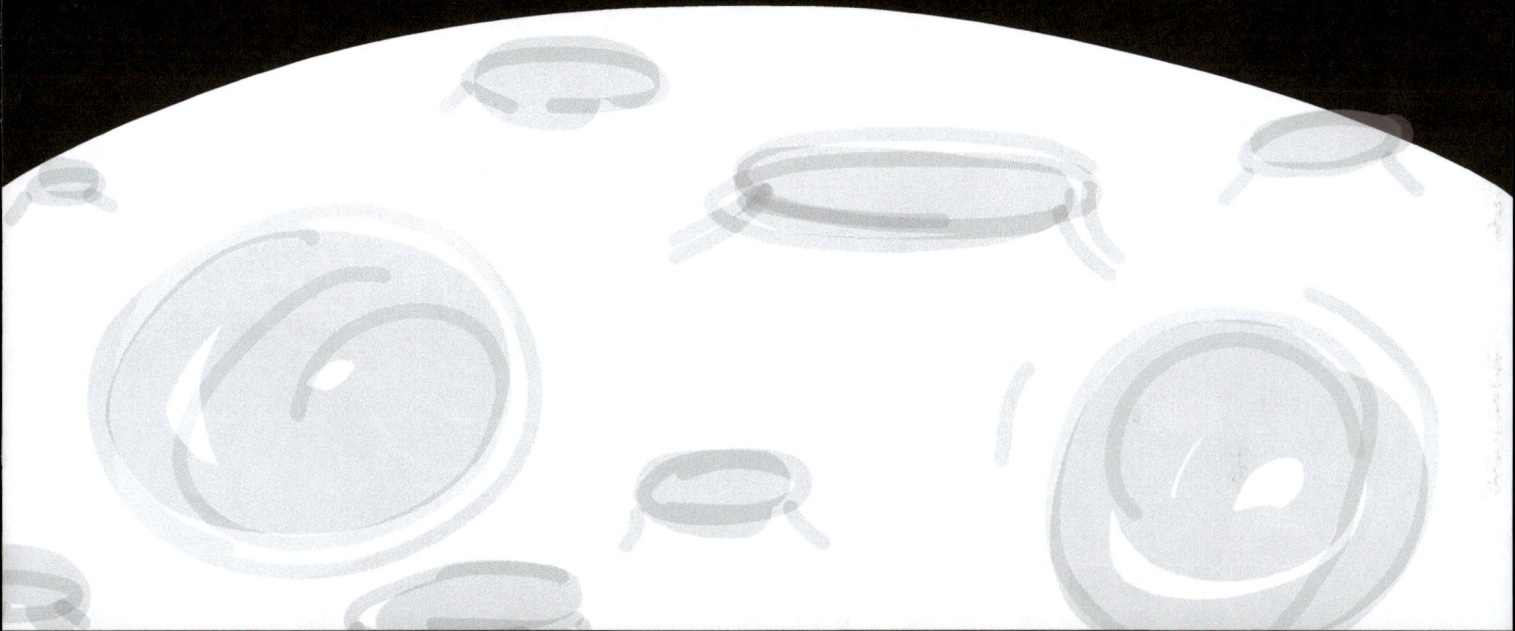

Sound it Out
1. KRA
2. TER

Atmosphere

Sound it Out
1. AT
2. MUS
3. FEER

The gases surrounding a planet.
Oxygen and nitrogen are gases in Earth's atmosphere.

Solar System

Neptune

Jupiter

Saturn

Venus

Uranus

Mercury

Earth

Mars

The Sun

All of this and more is in our Solar System.

The **Solar System** is the **Sun** and everything that **orbits** around it.

The **Sun** is the center of the **Solar System**.

What do you think **orbits** around the **Sun**?

Sun

Sound it Out

1. SUN

The **Sun** is a star.
A star is a big ball of hot gases that makes light and heat.

The **Sun** in our **Solar System** makes life on **Earth** possible.

What does it feel like when the **Sun** is out? What does it feel like when the **Sun** is behind a cloud?

The **Sun** is a very common type of star in the **Milky Way Galaxy**.

Milky Way Galaxy

Sound it Out

1. MILK
2. EE

Sound it Out

1. WAY

Sound it Out

1. GAL
2. X
3. EE

Solar System

The Milky Way Galaxy is VERY BIG! Do you see the bright stars in the sky at night? ALL of those stars are in the Milky Way Galaxy!

Our Solar System is a *very small part* of the Milky Way Galaxy!

Mercury

The Sun

Sound it Out

1. **MUR**
2. **KU**
3. **REE**

Mercury

The first **planet** from the **Sun** is **Mercury**.

Mercury is 36 million miles from the **Sun**

Solar System: #1 MERCURY

Mercury is the smallest planet in the Solar System. It is only a little bigger than our moon. It also looks like our moon. It is a solid surface that is covered with craters.

Mercury's very thin atmosphere contains the elements hydrogen, oxygen, sodium and helium.

Elements are discussed in
Chemistry: The Atom and Elements,
Book 2 of the Super Smart Science Series™.

What is atmosphere?

What is a crater?

venus

The Sun

Sound it Out
1. VEE
2. NUS

Venus

The second planet
from the Sun is Venus.

Venus is 67 million miles from the Sun.

Venus is almost as big as **Earth**. It has a solid surface covered in mountains, **craters** and volcanoes.

The **atmosphere** around **Venus** contains the elements nitrogen and carbon dioxide.

Venus spins backwards compared to other **planets**!

What is the name of the first planet from the Sun?

Solar System: #3 EARTH

The Sun

Earth

Sound it Out

1. ERTH

Earth

The third planet
from the Sun is Earth.

Earth is 93 million miles from the Sun.

Earth is very special.
70% of Earth is covered with water,
making it different from other
planets in the Solar System.

Earth's atmosphere contains the
elements nitrogen and oxygen.

The land surface of Earth
has mountains, canyons,
valleys and flat land.

What do you know about Earth?

The Sun

Sound it Out

1. MARZ

Mars

**The fourth planet
from the Sun is Mars.**

Mars is 228 million miles from the Sun.

Mars is like a cold desert.
Mars is similar to Earth
because it has seasons, polar caps
and varied land surfaces..

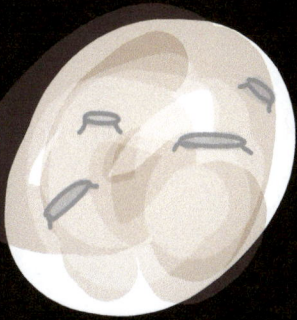

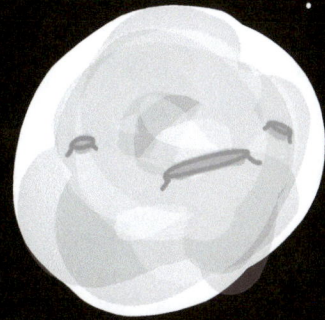

Mars has TWO moons!

The atmosphere around Mars
contains the elements
carbon dioxide and nitrogen.

What are the first
3 planets from the Sun?

Which planet is
covered mostly by water?

Jupiter

The Sun

Sound it Out

1. JU
2. PEH
3. TER

Jupiter

The fifth planet from the Sun is Jupiter.

Jupiter is the BIGGEST planet in the Solar System.

Jupiter is 484 million miles from the Sun.

Jupiter has many moons, over 50!

There is a HUGE storm called the *Great Red Spot* that has existed for hundreds of years!
(It is so big that 3 Earth's can fit inside!)

Jupiter's atmosphere contains hydrogen and helium.

Jupiter is not like the first 4 planets because it is a gas-giant planet. It does NOT have a solid surface.

What are the first 5 planets from the Sun?

The Sun

Saturn

Sound it Out
1. SA
2. TURN

Saturn

The sixth planet
from the Sun is Saturn.

Saturn is 886 million miles from the Sun.

Saturn is unique because of its 7 fantastic rings that are made of ice and rock.

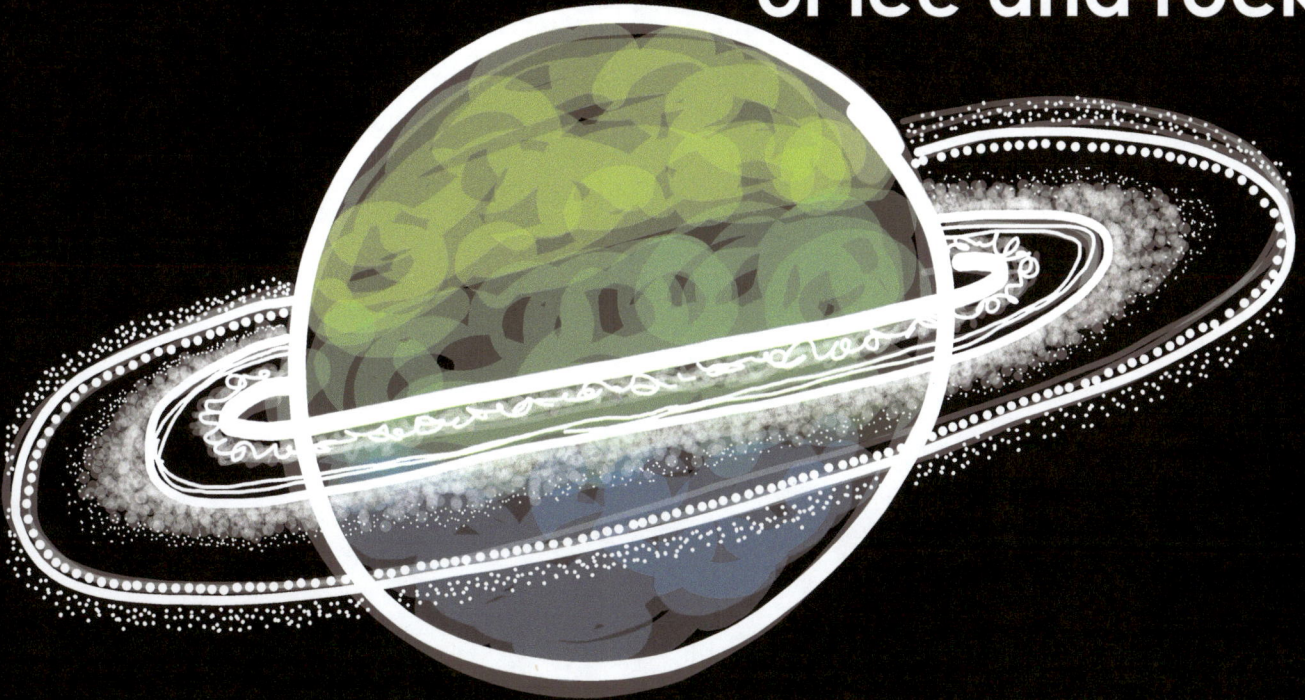

Saturn's **atmosphere** contains hydrogen and helium.

Saturn, like **Jupiter**, is a gas-giant and has over 50 moons.

What is the name of the HUGE storm on Jupiter?

The Sun

Uranus

Sound it Out

1. **UR**
2. **A**
3. **NUS**

Uranus

The seventh planet from the Sun is Uranus.

Uranus is 1.8 billion miles from the Sun.

A molecule called methane gives Uranus its blue color.

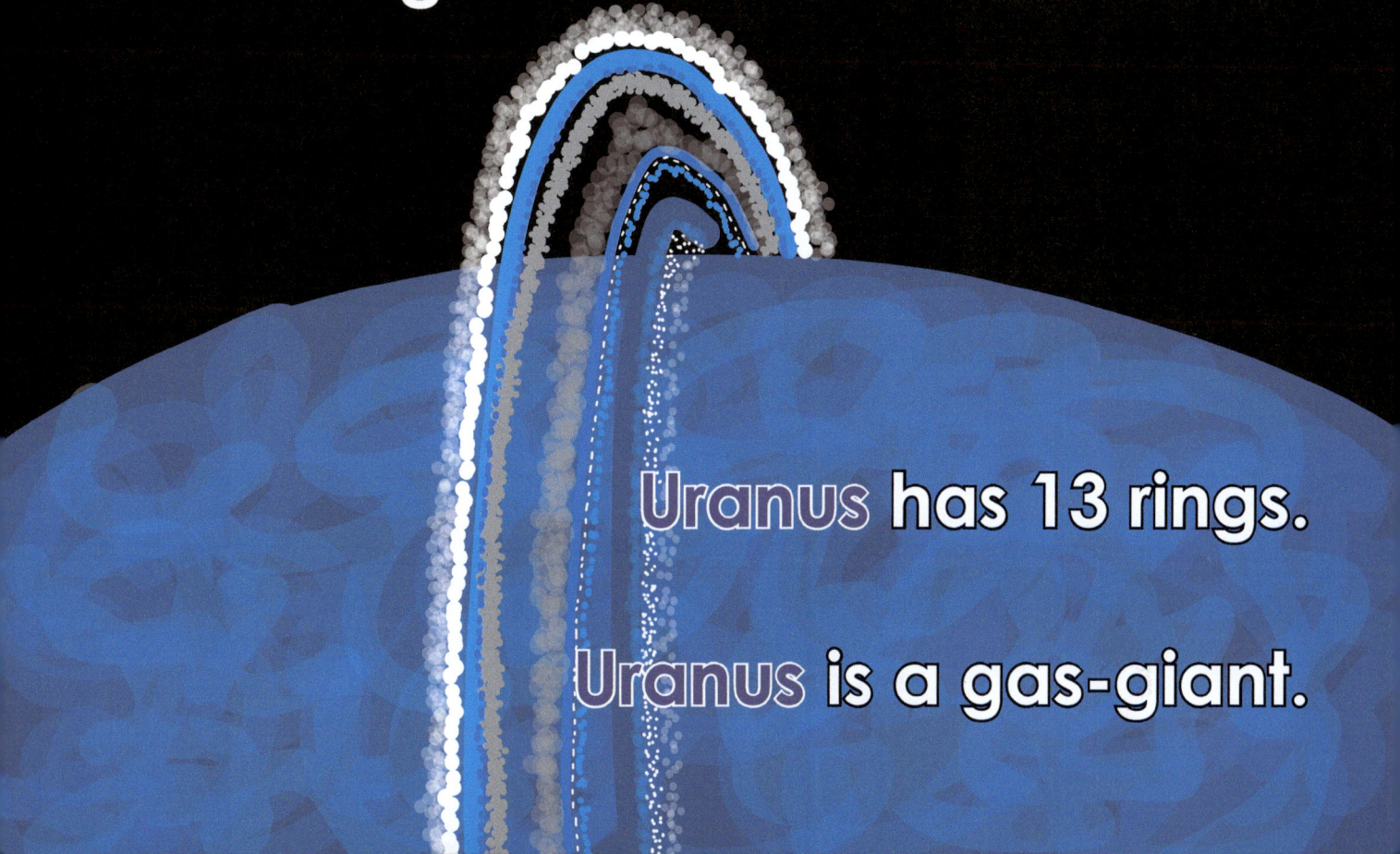

Uranus has 13 rings.

Uranus is a gas-giant.

The atmosphere of Uranus contains hydrogen and helium.

Uranus spins backwards!

What other planet spins backwards?

Neptune

The Sun

Sound it Out

1. NEP
2. TOON

Neptune

The eighth **planet** from the **Sun** is Neptune.

Neptune is 2.8 billion miles from the **Sun.**

Neptune is the farthest planet from the Sun in the Solar System.

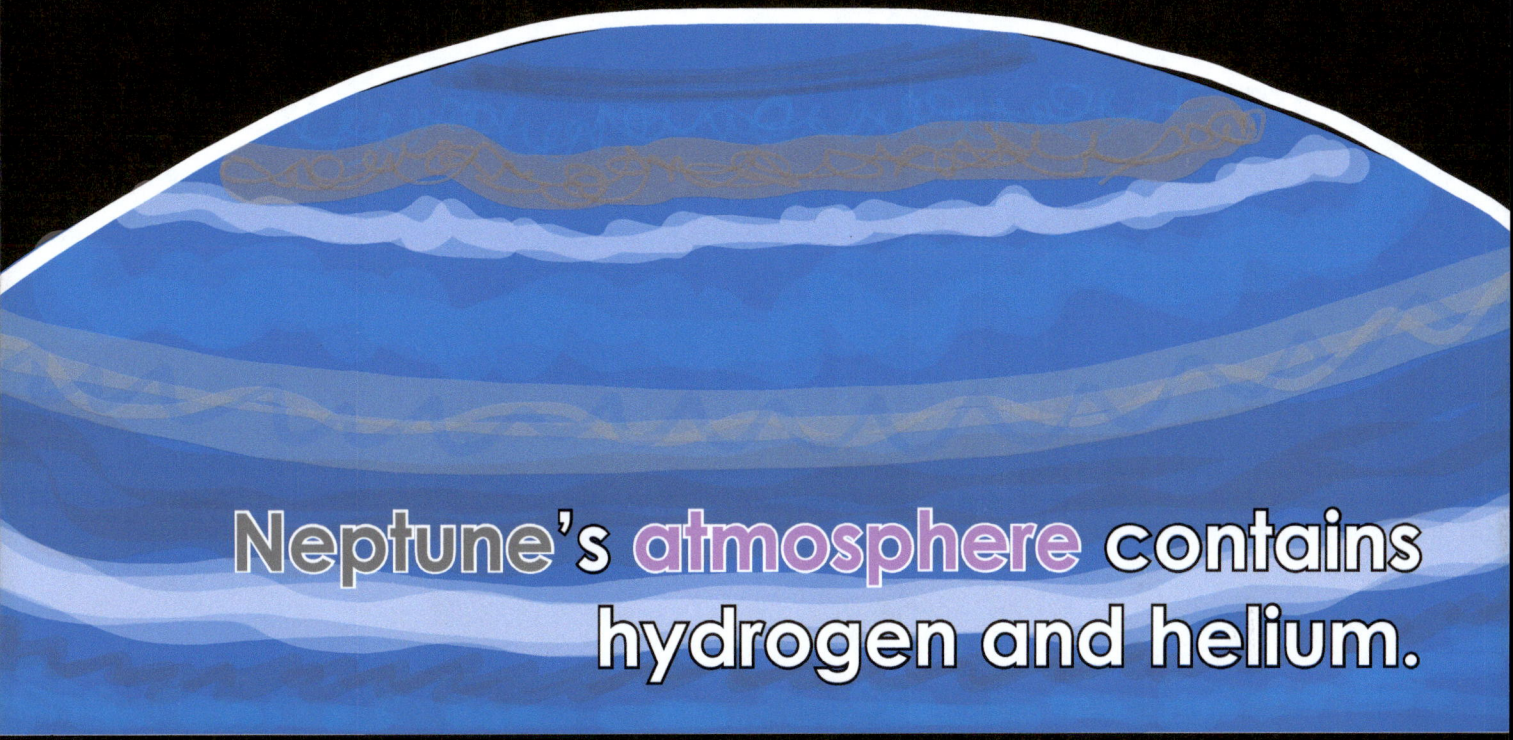

Neptune's atmosphere contains hydrogen and helium.

Neptune is a gas-giant.

What other planets are gas-giants?

Can you name all of the planets beginning closest to the Sun?

QUIZ

What planets are gas-giants?

What is the closest planet to the Sun in the Solar System?

What is the farthest planet from the Sun in the Solar System?

Which planets spin backwards?

What is the name of the HUGE storm on Jupiter?

The Solar System is in what Galaxy?

TRUE OR FALSE?

70% of Earth is covered by water.

Saturn has less than 5 moons.

Neptune has 13 rings.

The Sun is the center of the Solar System.

Uranus is blue because of oxygen in the atmosphere.

What else is in the Solar System other than planets and the Sun?

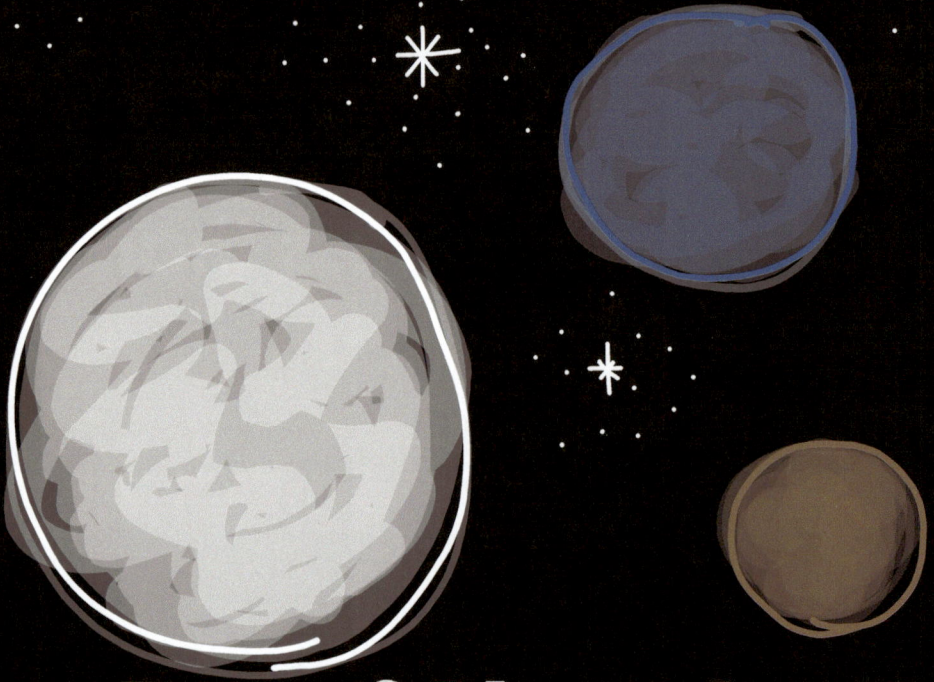

Dwarf Planet

Sound it Out

1. DWRF

Sound it Out

1. PLAN
2. IT

Dwarf planets orbit the Sun. Dwarf planets are much smaller than regular planets and can be found farther out past Neptune. Pluto is a dwarf planet.

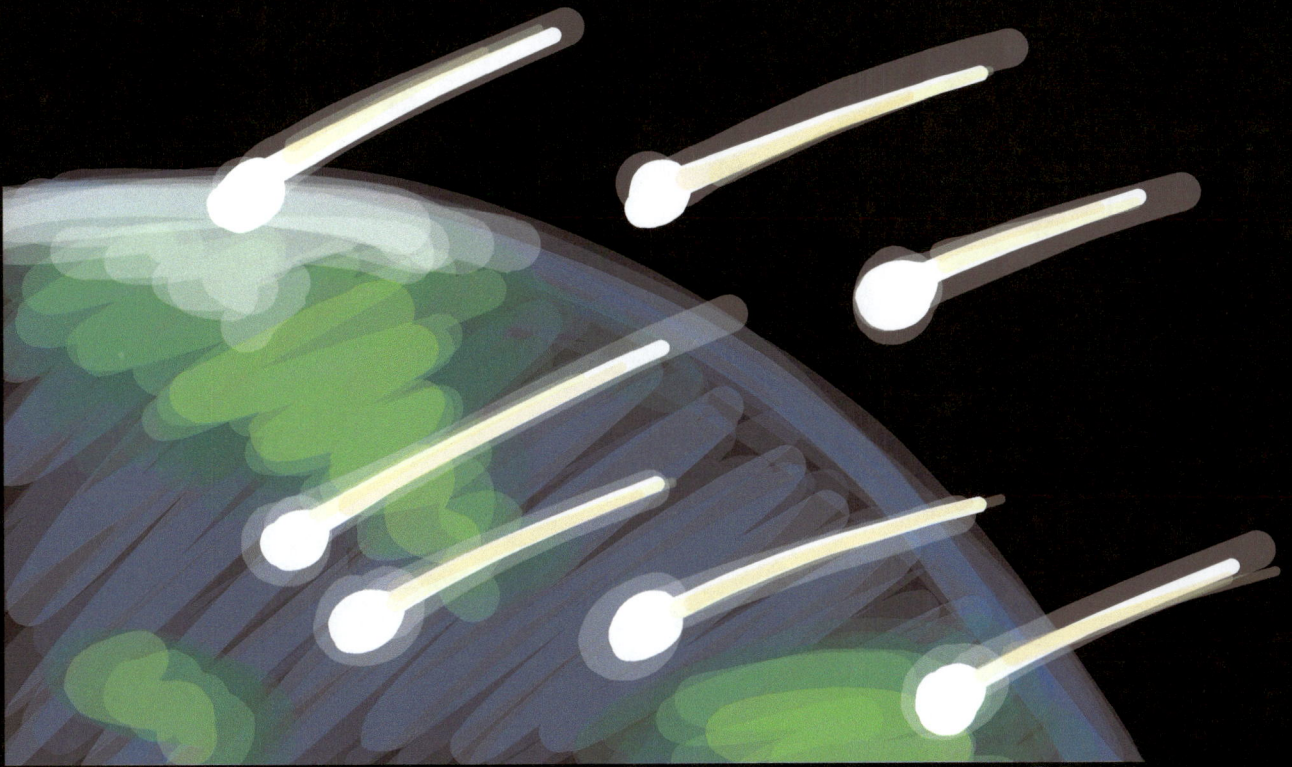

Meteors and Meteorites

Sound it Out

1. MEE
2. TEE
3. OR

Sound it Out

1. MEE
2. TEE
3. OR
4. ITE

Meteors are made of rock. As meteors enter Earth's atmosphere, they become shooting stars! Meteorites are meteors that go through the atmosphere and hit the land surface of a planet.

Asteroids

Asteroids are pieces of rock that orbit the Sun. There is a large area called the Asteroid Belt between Mars and Jupiter where there are millions of asteroids!

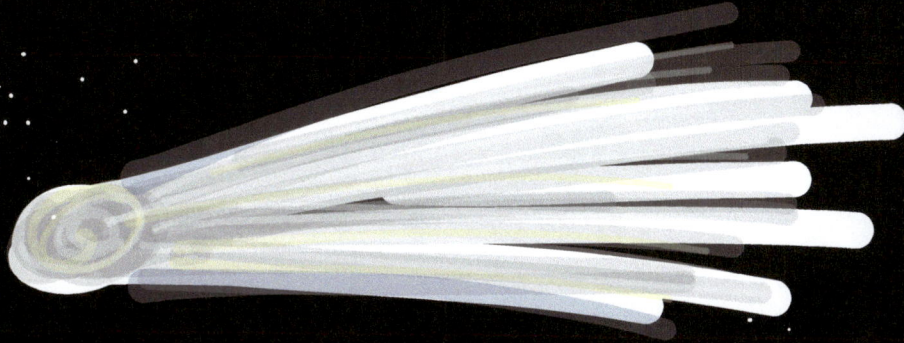

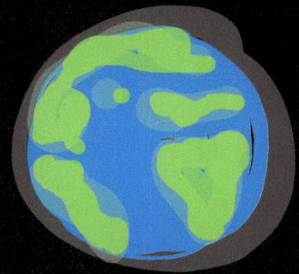

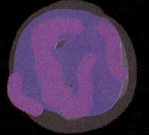

Comets

Sound it Out

1. COM
2. IT

Comets are big frozen balls of rock and gases. When a comet gets close to the Sun, it gets hot and creates a large glowing ball with a very long tail that you can sometimes see from Earth!

The Solar System is AMAZING!

A person who studies the Solar System is called an Astronomer.

Astronomer

Sound it Out
1. AH
2. STRON
3. O
4. MER

Telescope

Sound it Out
1. TEL
2. EH
3. SKOP

Satellite

Sound it Out
1. SAT
2. EL
3. LITE

Telescope

Satellite

Astronomers look through telescopes and see far out in space. They also monitor satellites that send information back to Earth.

Supernova

Sound it Out

1. SU
2. PER
3. NO
4. VA

Astronomers see very interesting things in space, like supernovas. A supernova is an exploding star!

They can even see farther out into the Milky Way Galaxy past the Solar System!

The Solar System is an exciting place where 8 planets orbit the Sun.

Each planet has elements in the atmosphere.

Some planets have a land surface, some planets are gas-giants.

In the Solar System there are also Dwarf Planets, Meteors, Asteroids and Comets.

The Solar System is only a small part of the Milky Way Galaxy.

Astronomers look through telescopes and use satellites to explore stars, planets and supernovas everyday!

Review the terms you learned!

Solar System

Sun

Planet

Crater

Atmosphere

Orbit

Mercury

Venus

Earth

Mars

Dwarf Planet

Asteroid

Jupiter

Supernova

Telescope

Saturn

Uranus

Neptune

Satellite

Astronomer

Milky Way Galaxy

Meteor

Comet

Anatomy & Physiology #1

Bones, Muscles, and The Stuff That Connects Bones and Muscles

Ages 0 - 100

Skull

Clavicle

Vertebrae

Sternum

Humerus

Radius

Rib

Ulna

Pelvis

Phalanges

Femur

Fibula

Tibia

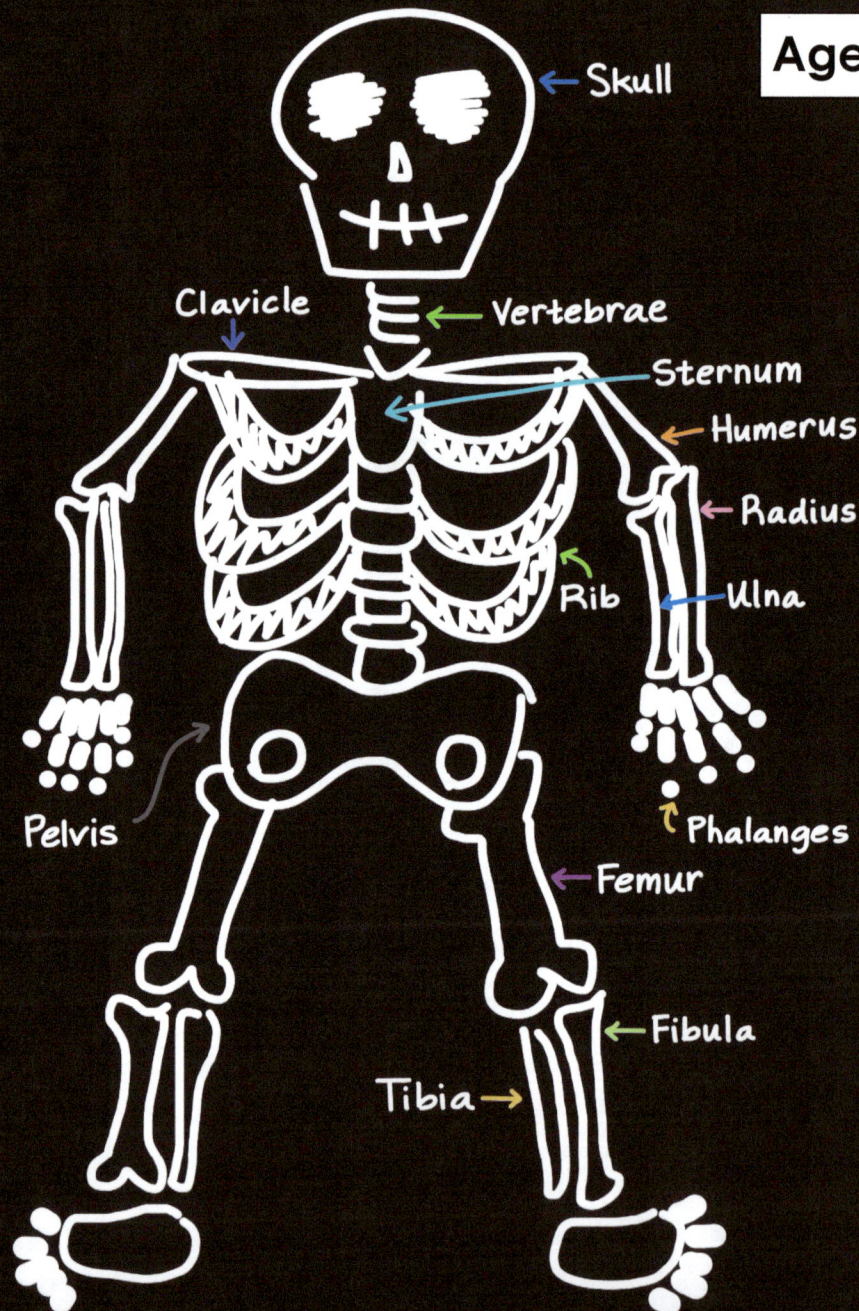

Super Smart Science Series - Book #5

By: April Chloe Terrazas

Anatomy & Physiology

1

Bones Muscles and The Stuff That Connects Bones and Muscles

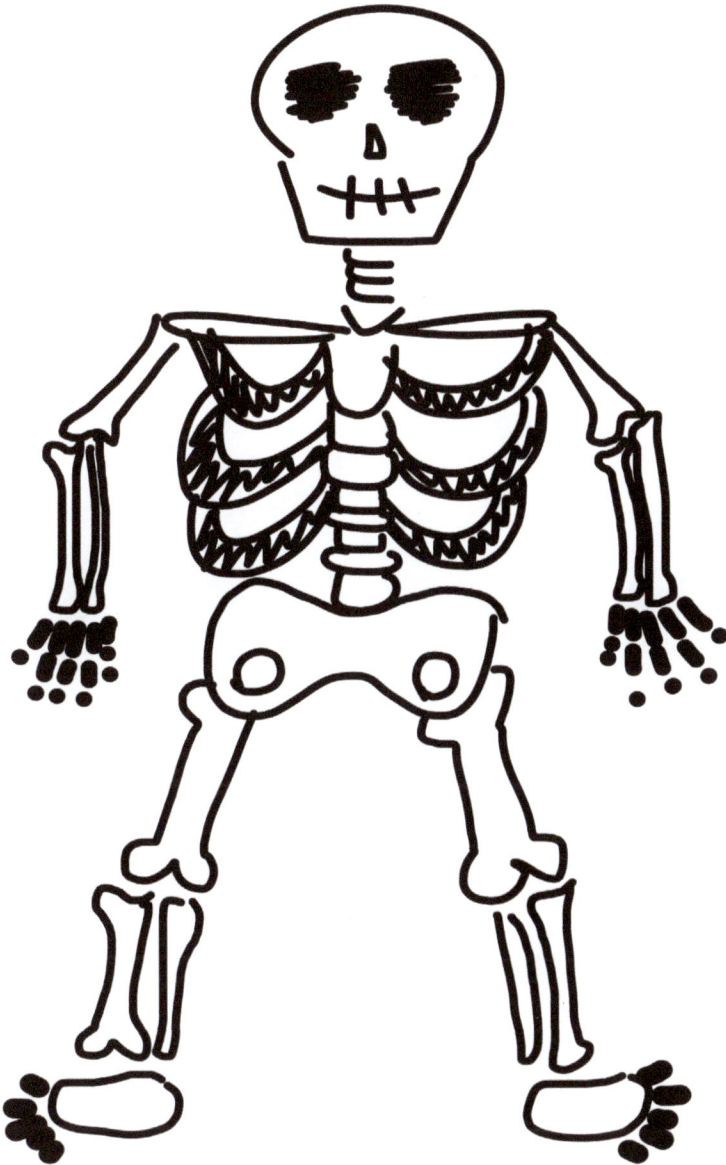

This book is dedicated to my Mom & Dad, Kim and Joel Terrazas.

Thank you so much for your wholehearted support of all of my endeavors since early childhood!

By: April Chloe Terrazas

Anatomy & Physiology PART 1: Bones, Muscles, and The Stuff That Connects Bones and Muscles.
April Chloe Terrazas, BS University of Texas at Austin.
Copyright © 2014 Crazy Brainz, LLC

Visit us on the web! www.Crazy-Brainz.com

Cover design, illustrations and text by: April Chloe Terrazas

Anatomy art competition winners:
Taj Estrada and Sydney Estrada!

Congratulations!
We LOVE your portrayal of
human anatomy!

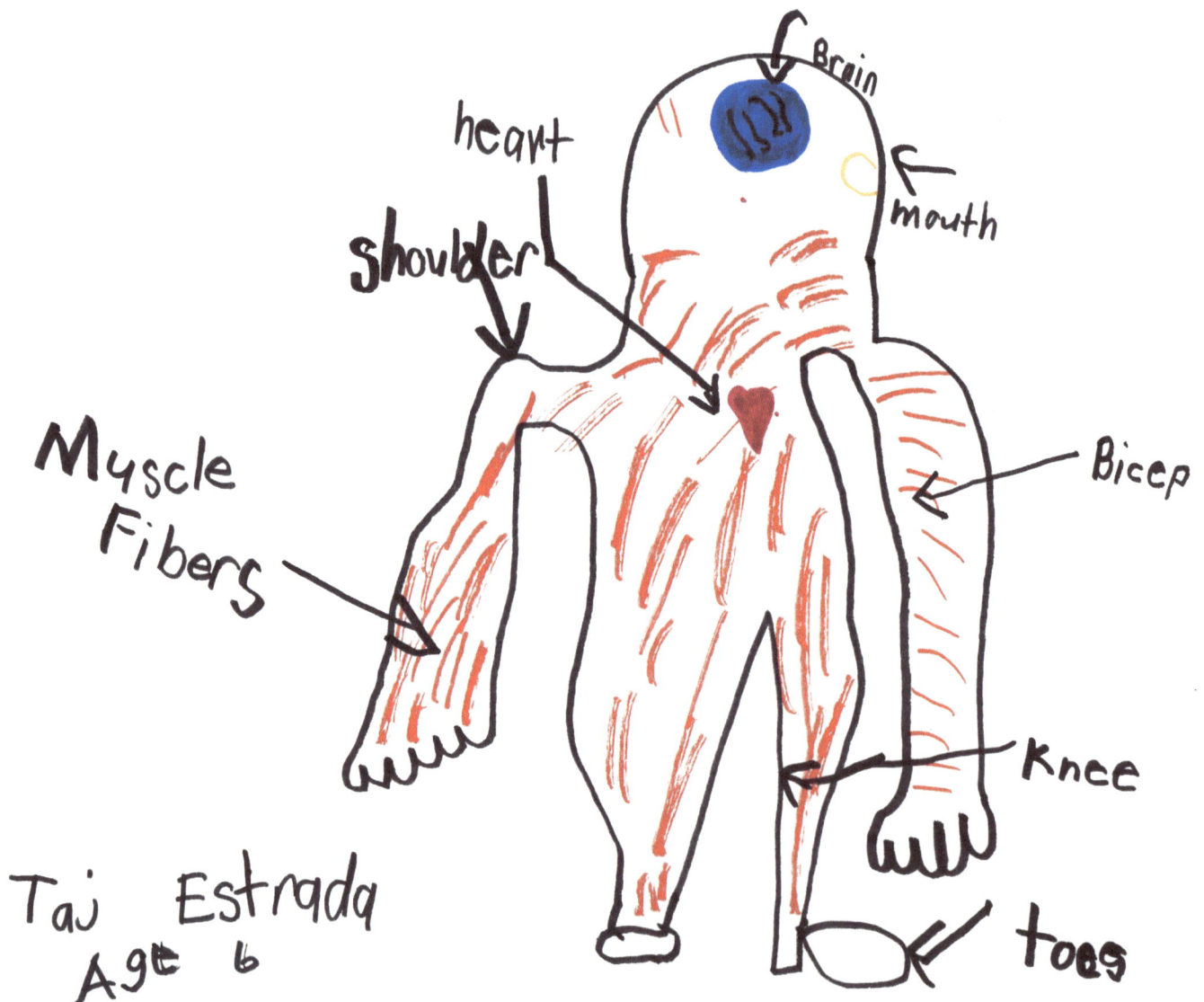

Brain

heart

mouth

shoulder

Muscle
Fibers

Bicep

Knee

Taj Estrada
Age 6

toes

The Inside of a Bone

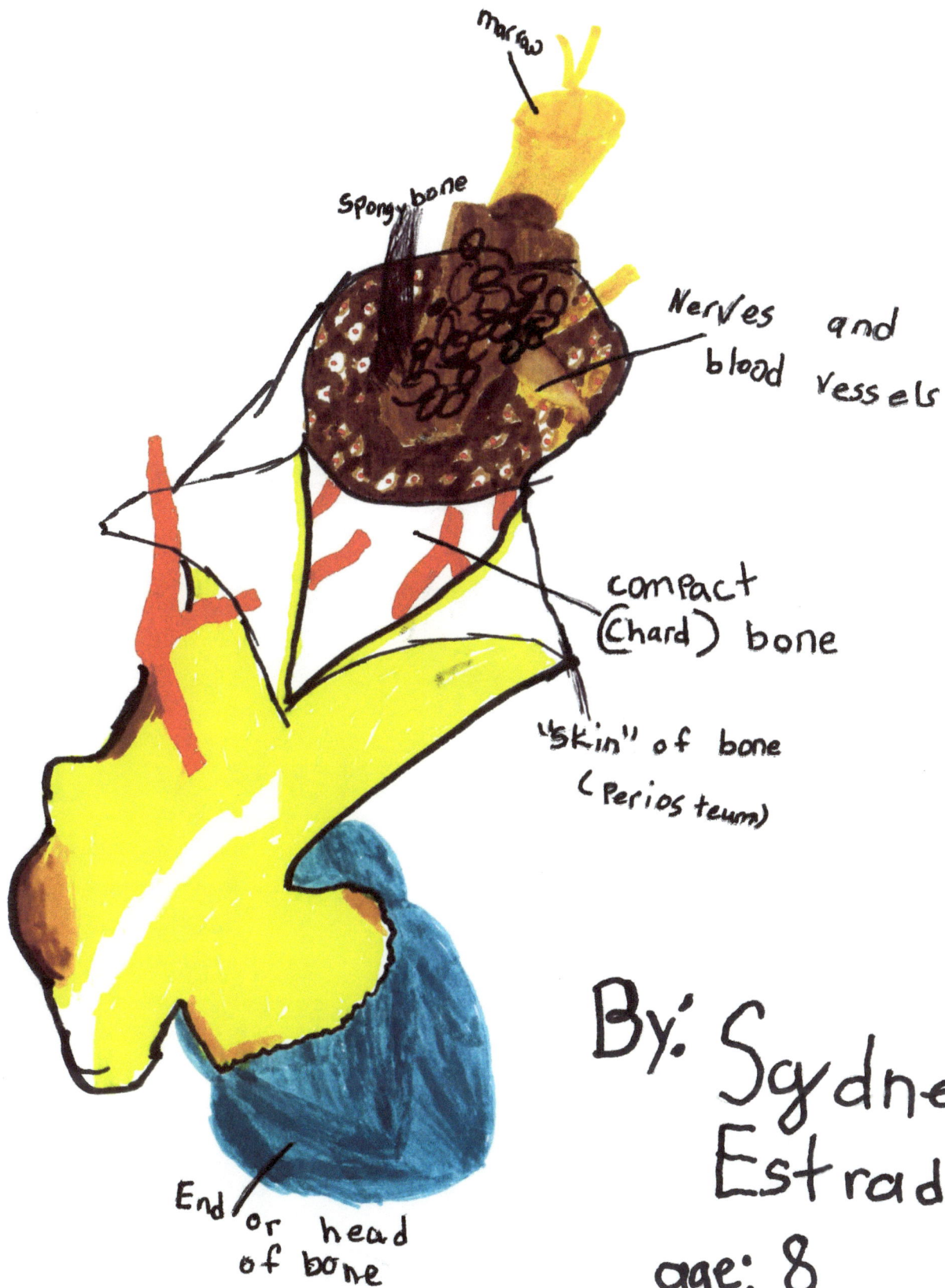

marrow

Spongy bone

Nerves and blood vessels

compact (hard) bone

"skin" of bone (Periosteum)

End or head of bone

By: Sydney Estrada

age: 8

Anatomy

Sound it Out

1. UH
2. NA
3. TO
4. ME

Physiology

Sound it Out

1. FIZ
2. E
3. OL
4. O
5. JEE

Anatomy is the structure of a living organism *(how it looks)*. For example, the hand is small with fingers on it.

Physiology is the function of a living organism, *(how it works)*. For example, the hand moves to write or play an instrument.

What is the **anatomy** of your face?

How does it look?

Eyes
Nose
Mouth

What is the **physiology** of your face?

How does it work?

See
Smell
Talk
Eat

Bone

Sound it Out

1. BONE

Skeleton

Sound it Out

1. SKEL
2. EH
3. TUN

Bones are hard tissue that make up the **skeleton**. Without **bones**, we would be like a ball of jelly!

Bones are alive and can grow and change shape.

A human adult has 206 bones!

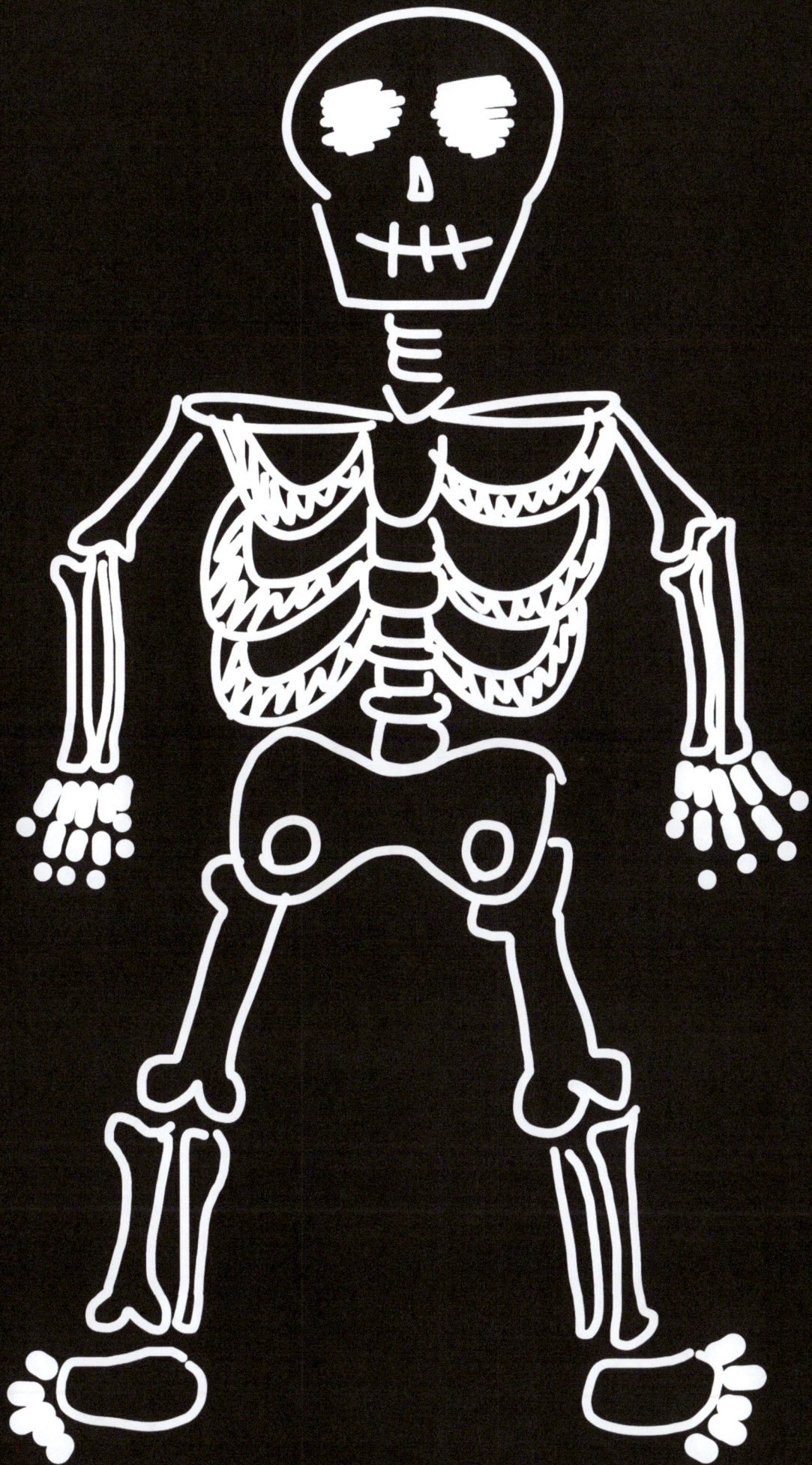

This is a <u>long bone</u>.
<u>Long bones</u> are in your legs, arms and even your fingers!

The epiphysis is the end of a <u>long bone</u>.

The diaphysis is the shaft of a <u>long bone</u>.

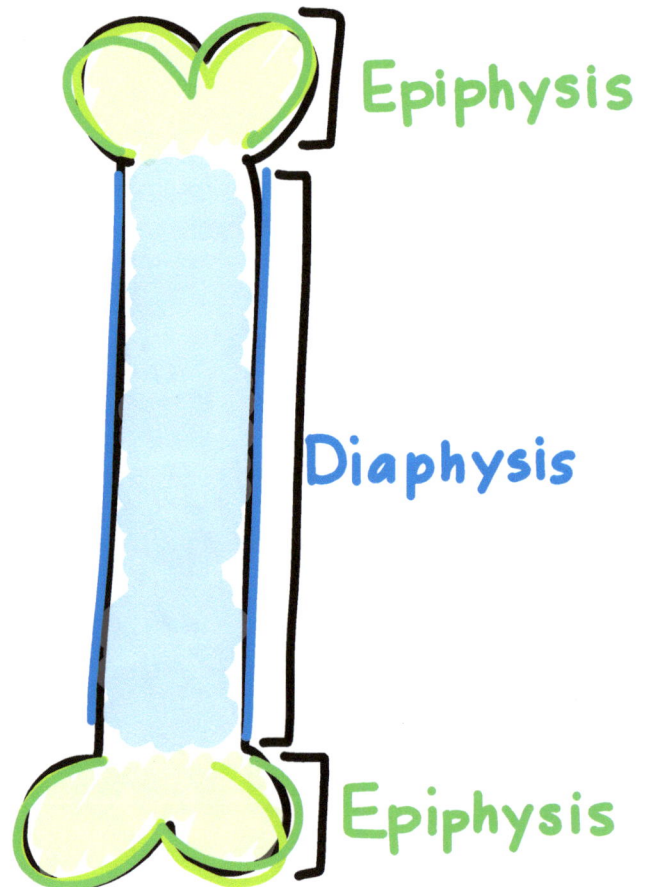

Epiphysis

Diaphysis

Epiphysis

Epiphysis

Sound it Out

1. E
2. PIF
3. EH
4. SUS

Diaphysis

Sound it Out

1. DI
2. AF
3. EH
4. SUS

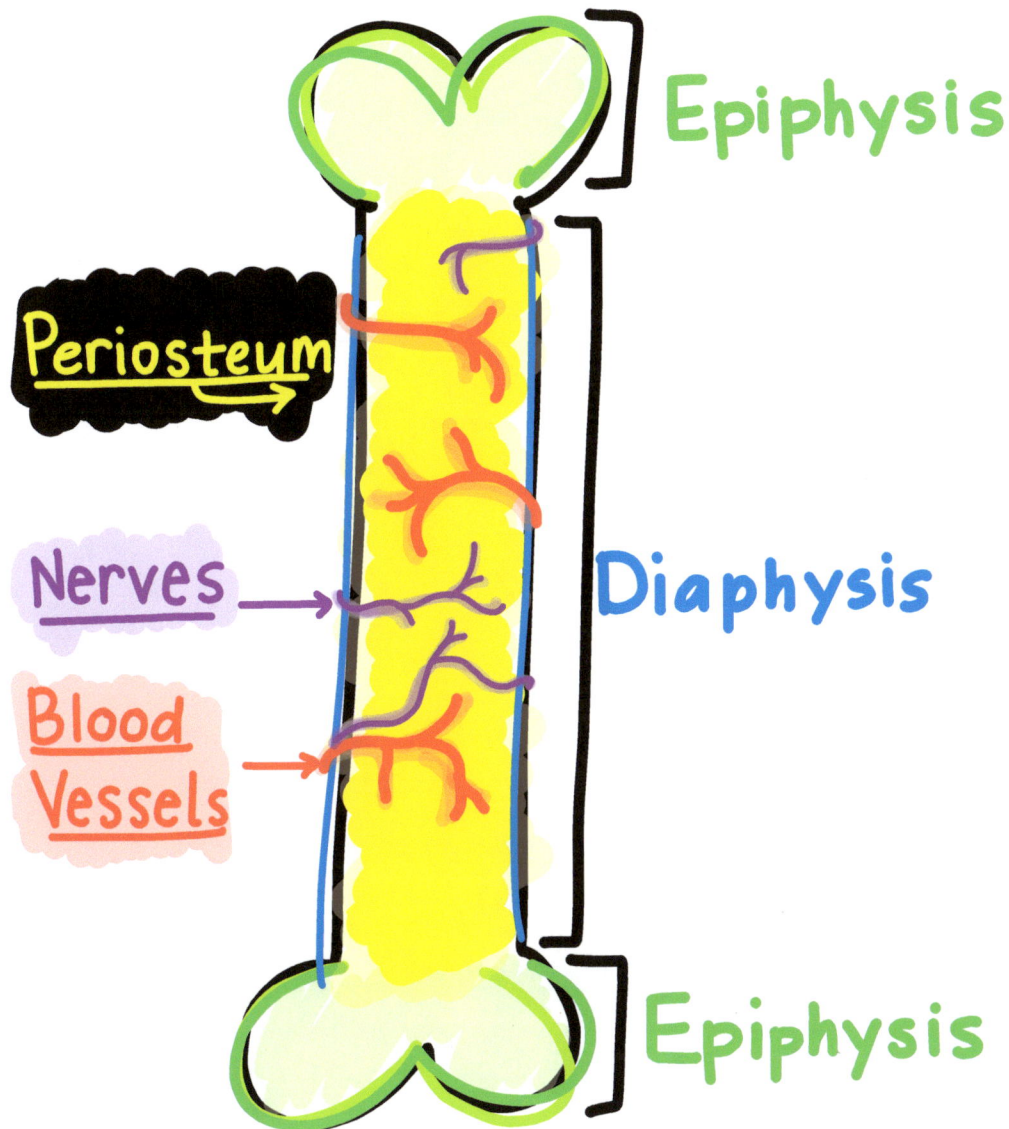

Epiphysis

Periosteum

Nerves

Blood Vessels

Diaphysis

Epiphysis

Periosteum is a membrane that covers the outer surface of bones and has nerves and blood vessels.

Periosteum

Sound it Out

1. PER
2. EE
3. OS
4. TEE
5. UM

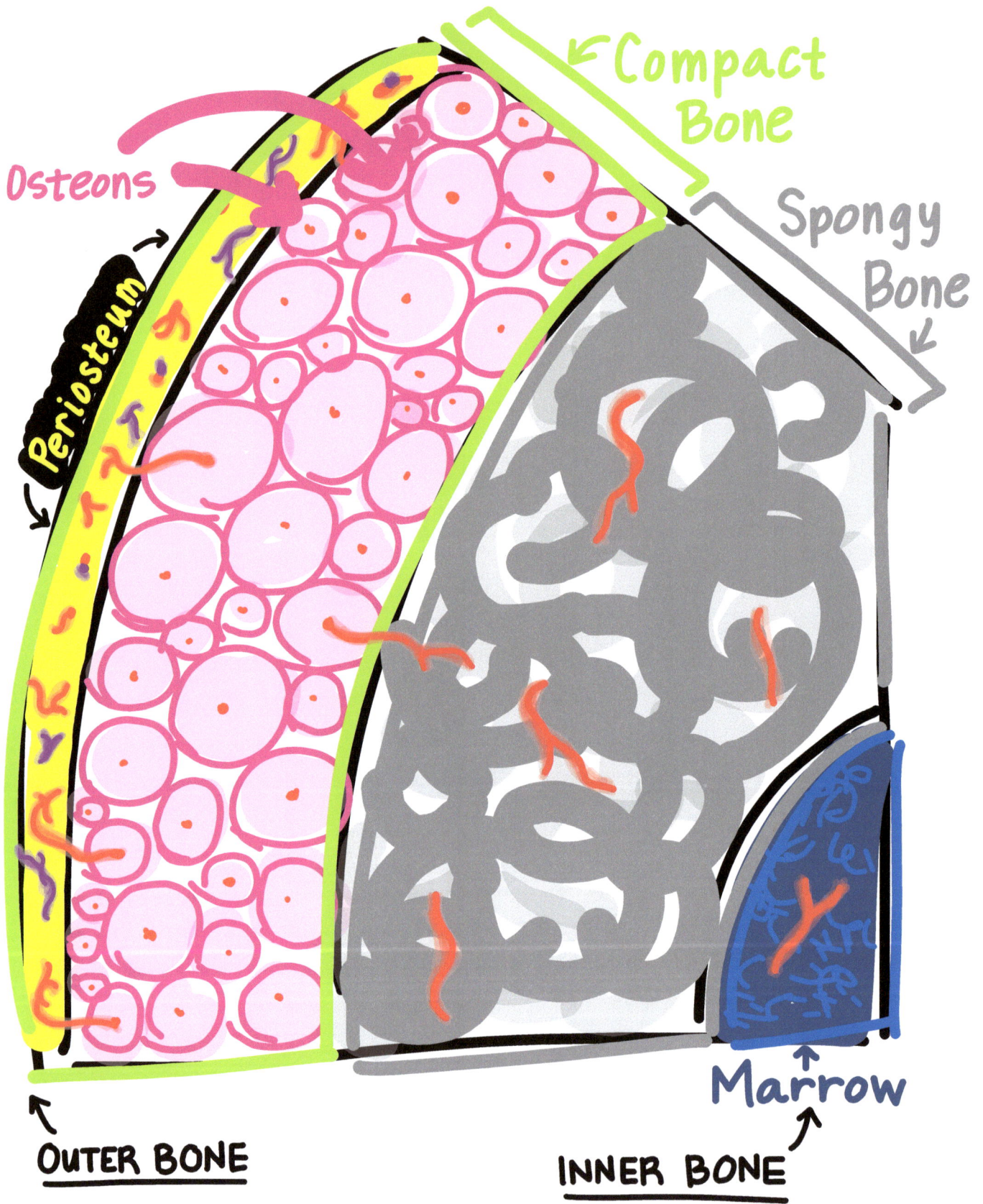

Osteons

Periosteum

← Compact Bone

Spongy Bone ↓

OUTER BONE

INNER BONE

Marrow ↑

(Inside a bone)

The outer layer of the bone is called compact bone.

Compact bone is made of osteons and periosteum.

Osteons are layers of dense bone with a canal inside for blood vessels.

Do you see the red blood vessels inside the osteons?

Osteon

Compact Bone

Sound it Out

1. OS
2. TEE
3. ON

Sound it Out

1. KOM
2. PAKT
1. BONE

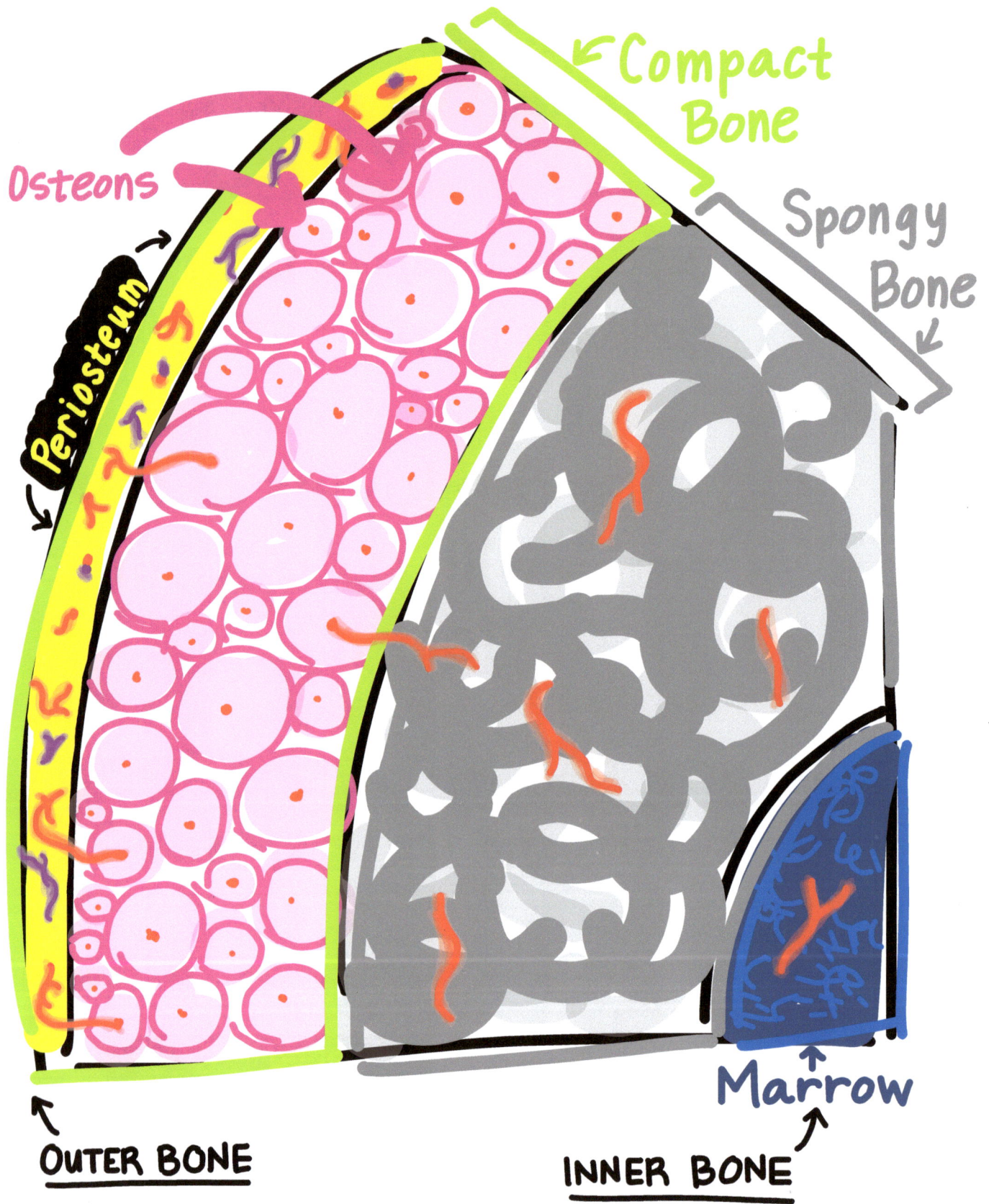

Osteons

Periosteum

Compact Bone

Spongy Bone

Marrow

OUTER BONE

INNER BONE

(Inside a bone)

The inner layer (the middle) of the bone is made of marrow and spongy bone.

Marrow is a soft substance that fills the spongy bone and makes blood cells.

What are osteons?
What is compact bone made of?
What is periosteum?

Spongy Bone

Marrow

Sound it Out
1. MER
2. O

Sound it Out
1. SPUN
2. JEE
1. BONE

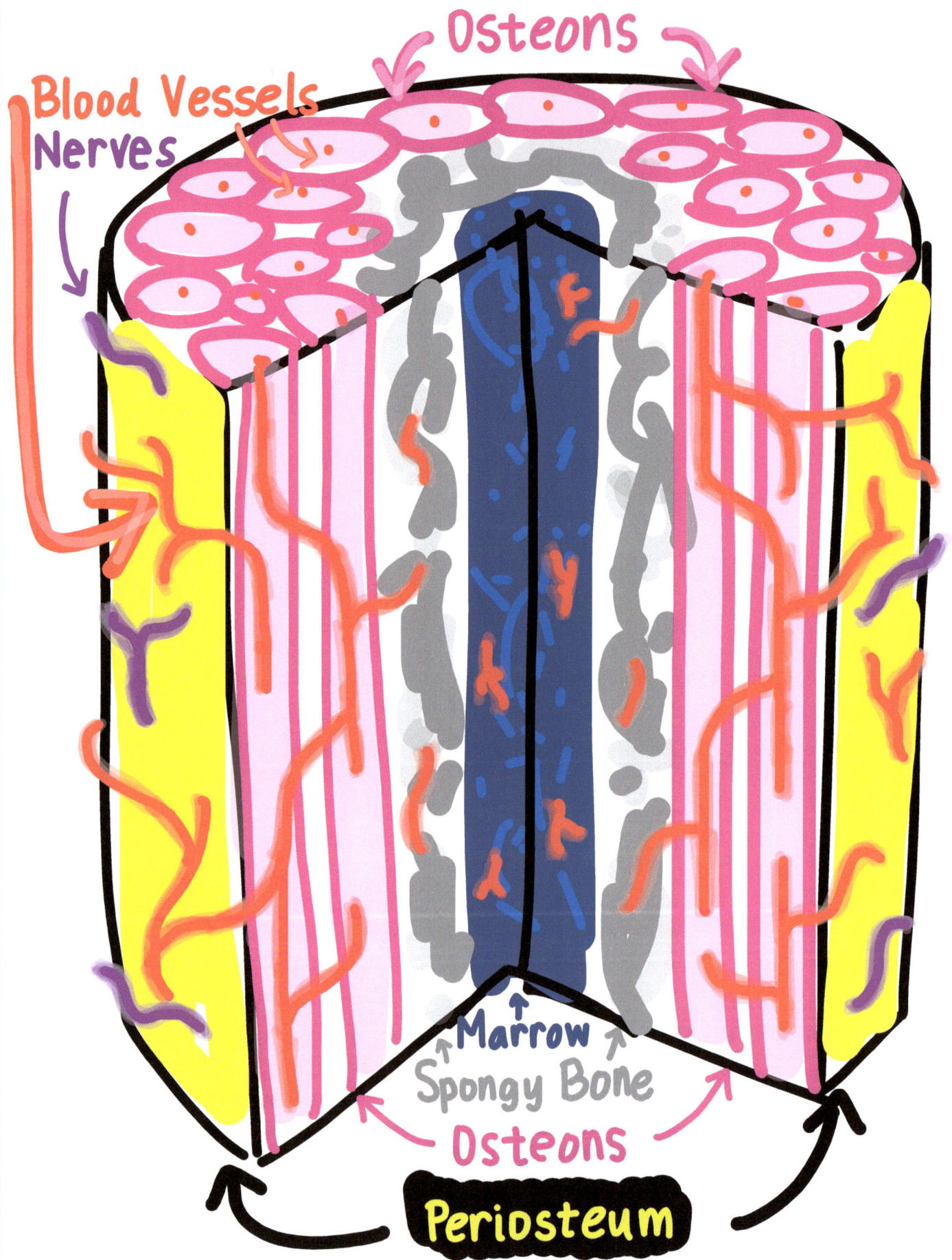

Osteons

Blood Vessels
Nerves

Marrow
Spongy Bone
Osteons

Periosteum

Starting from the middle of the bone going outward:

Marrow

Spongy Bone

Osteons

Periosteum

Blood vessels go from the outer membrane of the periosteum into the bone!

This is how your bone heals when it is fractured or broken.

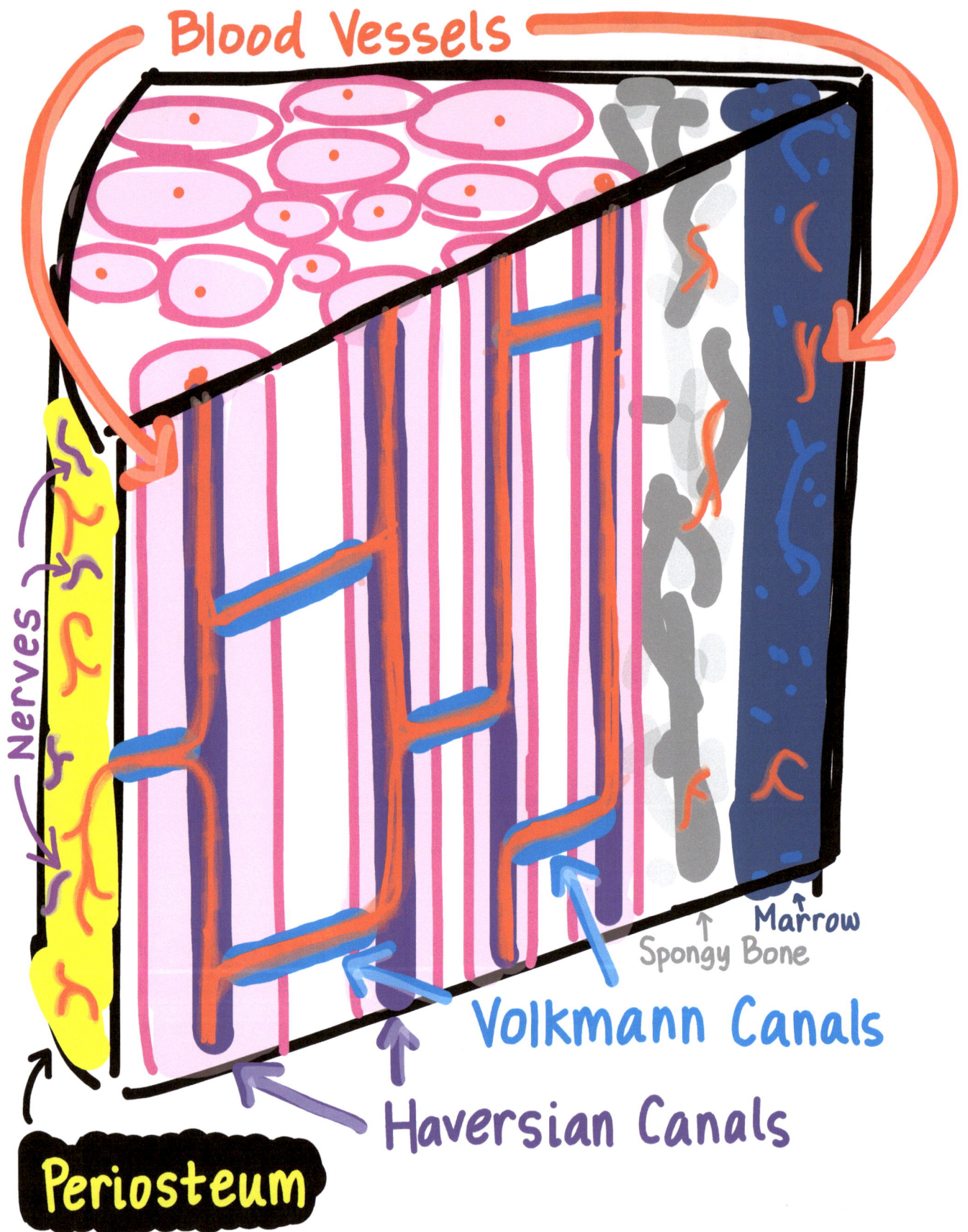

Blood Vessels

Nerves

Periosteum

Volkmann Canals

Haversian Canals

Spongy Bone

Marrow

Blood vessels are located in the canals of the osteons.

The canals inside the osteons are called Haversian Canals. They go *up and down*.

Haversian Canal

Sound it Out

1. HUH
2. VER
3. SHUN

1. KUH
2. NAL

Volkmann Canal

Sound it Out

1. VOLK
2. MUN

1. KUH
2. NAL

The blood vessels move from *side to side* between osteons through Volkmann canals.

Volkmann canals also connect the outer periosteum to the inner osteons.

The outer compact bone is
made of periosteum
and osteons.

The inner bone is made of
marrow inside spongy bone.

Blood vessels go from the
outer periosteum to the osteons
through volkmann canals that
go from *side to side*.

Blood vessels move
up and down the osteons
through haversian canals.

Blood vessels inside the bone
are the way fractured
or broken bones heal.

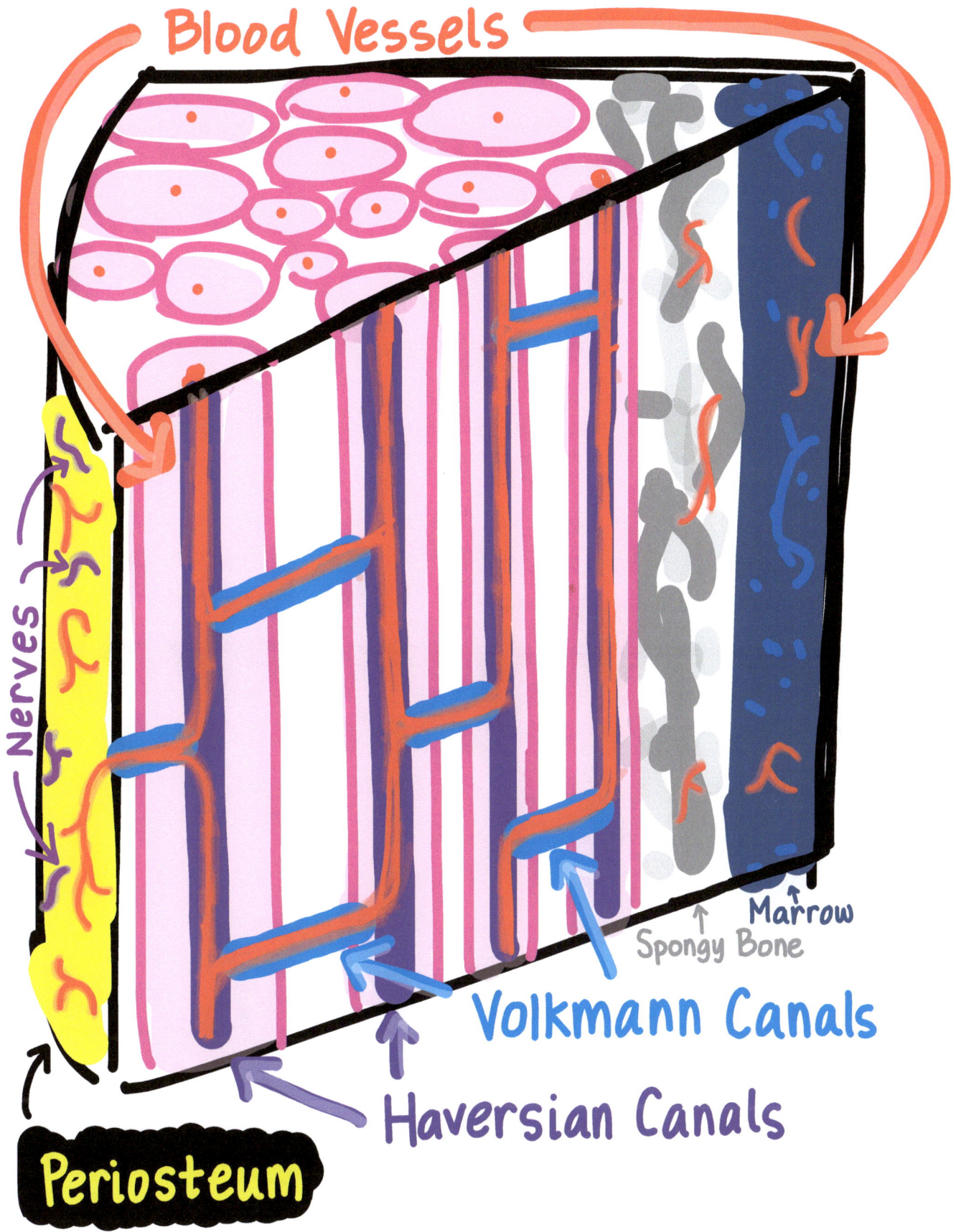

Blood Vessels

Nerves

Periosteum

Haversian Canals

Volkmann Canals

Spongy Bone

Marrow

What is the periosteum?

What is the difference between the haversian canals and the volkmann canals?

What is compact bone?

What is the difference between the epiphysis and diaphysis?

CONGRATULATIONS!

You are now a bone anatomy expert!

MUSCLE

There are 3 types of muscle in your body:

Skeletal
SKEL-EH-TUL

Cardiac
KAR-DEE-AK

Smooth
SMOOTH

Skeletal muscles are <u>voluntary</u> muscles. This means that <u>you control</u> the movement.

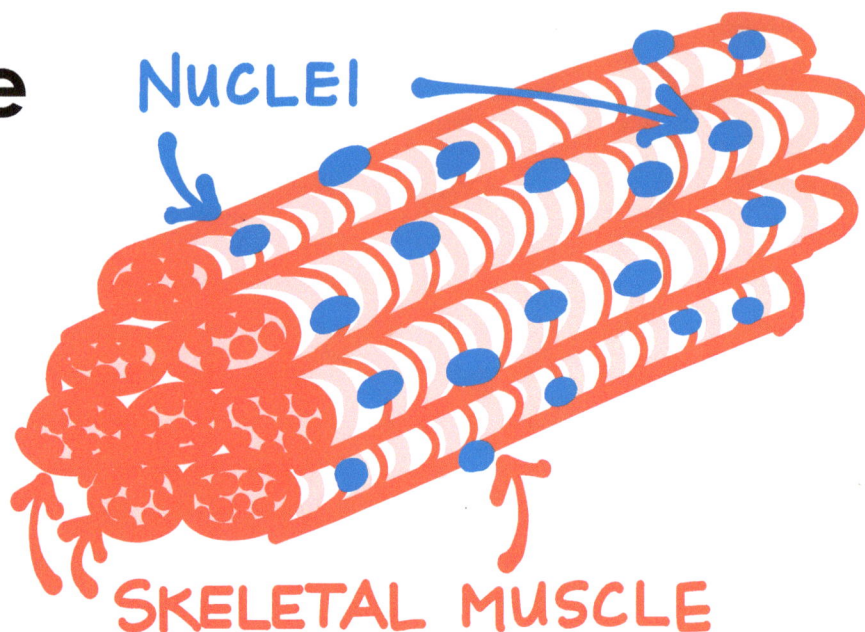

NUCLEI

SKELETAL MUSCLE

Skeletal muscles are striated. Striated means striped.

Do you see the striated skeletal muscle?

Skeletal muscles work together with the skeleton to move your body.

Striated

Sound it Out

1. STRI

2. A

3. TED

Skeletal muscles are attached to bones by tendons.

Tendons are made of white fiber-like tissue.

Tendons are VERY strong. They have to be because they hold the skeleton together.

How many bones are in a human skeleton?

Tendons allow you to run, walk, jump, lift and dance.

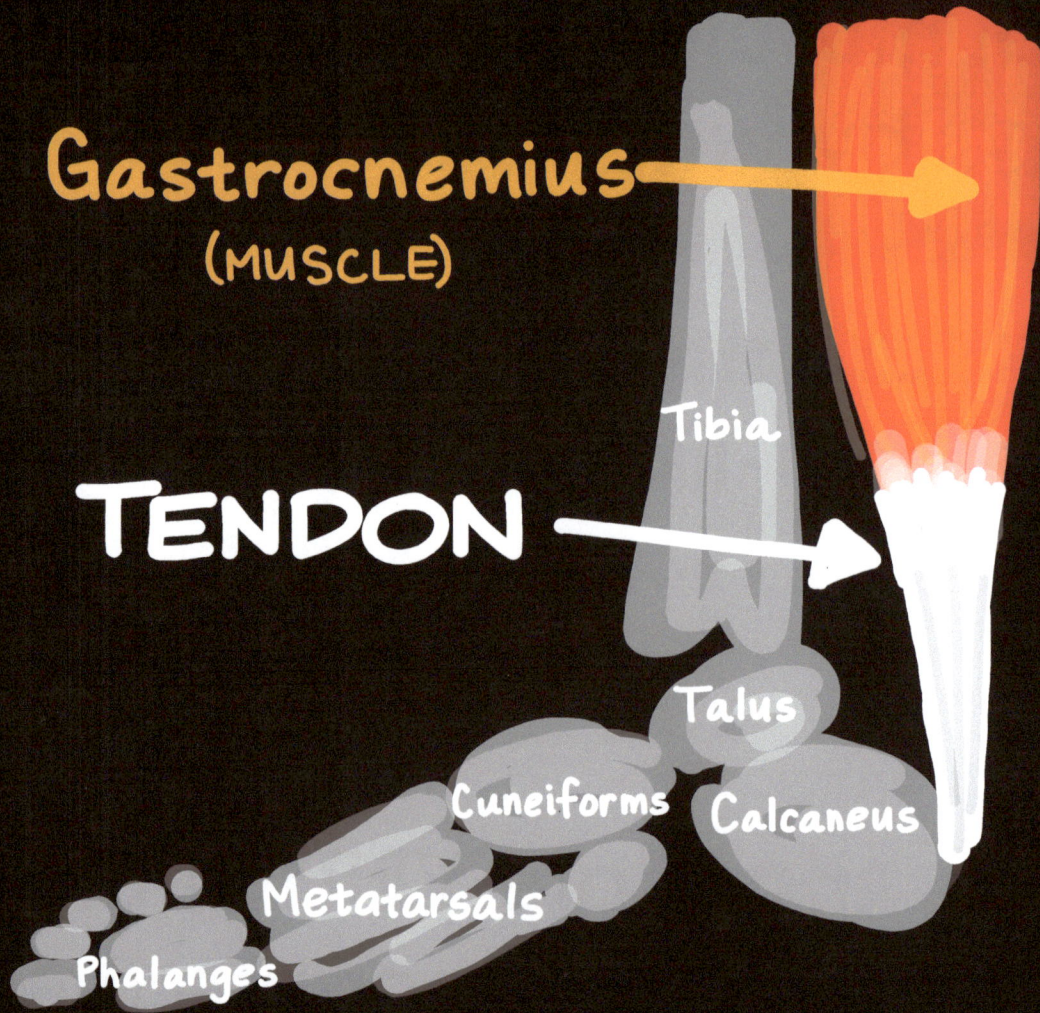

Gastrocnemius
(MUSCLE)

TENDON

Tibia

Talus

Cuneiforms

Calcaneus

Metatarsals

Phalanges

Gastrocnemius

This is a **tendon** that connects the heel to the **gastrocnemius** muscle so you can run and jump!

Sound it Out

1. GAS
2. TROK
3. NEE
4. MEE
5. US

Say the names of the bones in the lower leg and foot.

Tibia = TIB-EE-UH

Talus = TAL-US

Cuneiforms = Q-NEE-UH-FORMS

Calcaneus = KAL-KAN-EE-US

Metatarsals = MET-UH-TAR-SULS

Phalanges = FUH-LAN-JEES

The **tendon** conne
the **calcaneus bone**
gastrocnemius (skeleta

Tendons connect muscle to bone.

Ligaments connect bone to bone.

Ligament

Sound it Out
1. LIG
2. UH
3. MENT

Ligaments connect the femur (thigh bone) to the tibia (shin bone) at the knee joint.

Cartilage is in your joints, ears, nose and throat!

Cartilage

Sound it Out
1. KAR
2. TEH
3. LEJ

The **patella** is the knee cap.

Femur

Patella

← LIGAMENT

← Cartilage

Tibia

Cartilage provides cushion for this knee joint.

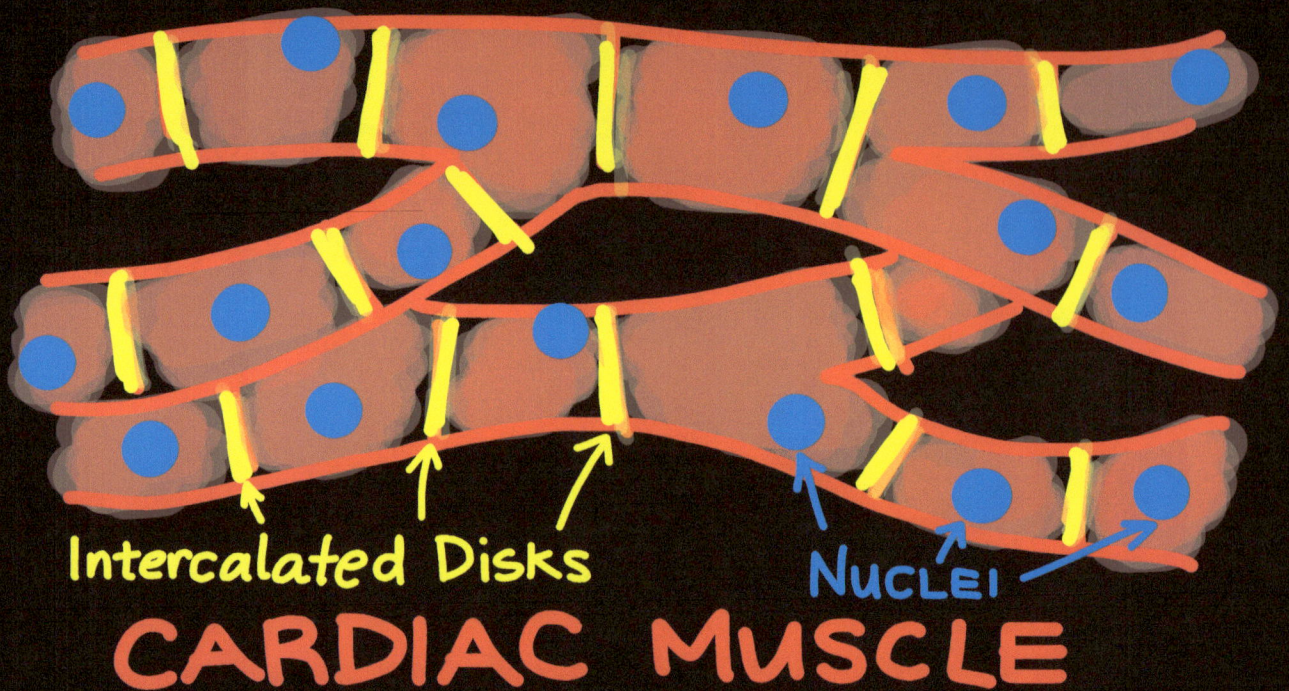

Intercalated Disks — Nuclei
CARDIAC MUSCLE

Cardiac muscle is <u>involuntary</u>.

<u>You do not control</u> your heart beat.

Cardiac muscle does not get tired. Cardiac muscle contracts to push blood out of the heart and relaxes to draw blood into the heart.

Cardiac muscle is ONLY in the heart.

Cardiac muscle cells are joined at intercalated disks.

Intercalated disks create the striations in cardiac muscle.

Intercalated

Sound it Out

1. IN
2. TER
3. KUH
4. LA
5. TED

Disks

Sound it Out

1. DSKS

NUCLEI

Smooth Muscle

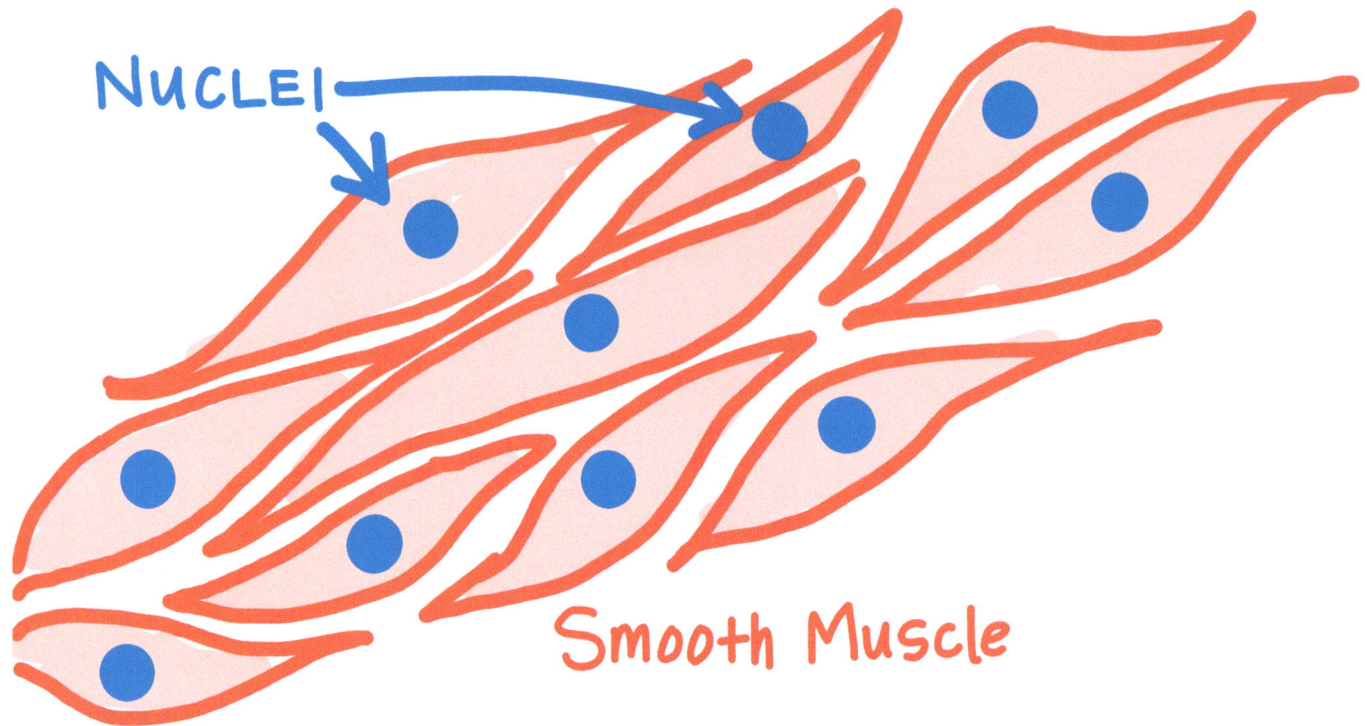

Smooth muscle is <u>involuntary</u>, it works automatically in your body.

Smooth muscle is in your stomach, intestines and even in your eye!

REVIEW:

Skeletal muscle is <u>voluntary</u>, striated and it is found all over your body.

Skeletal muscle connects to bones through tendons, like the calcaneus to the gastrocnemius.

Ligaments connect bone to bone, like the femur to the tibia.

Cardiac muscle is <u>involuntary</u> and found only in the heart.

Intercalated disks make the striations in cardiac muscle.

Smooth muscle is <u>involuntary</u> and found in your eye.

We know what bones and muscles look like, we know their anatomy. But how do our bones and muscles move?

When the muscle moves, the bones move because bones are connected to muscle through tendons.

What makes our muscles move?

The brain!

Neurons carry messages from our brain to our muscles.

(Review the structure and function of the neuron in Neurology: The Amazing Central Nervous System - book 3 of the Series - and elements in Chemistry: The Atom and Elements - book 2 of the Series)

The message
to move is sent
from the brain,
through the
neurons, to the
muscle.

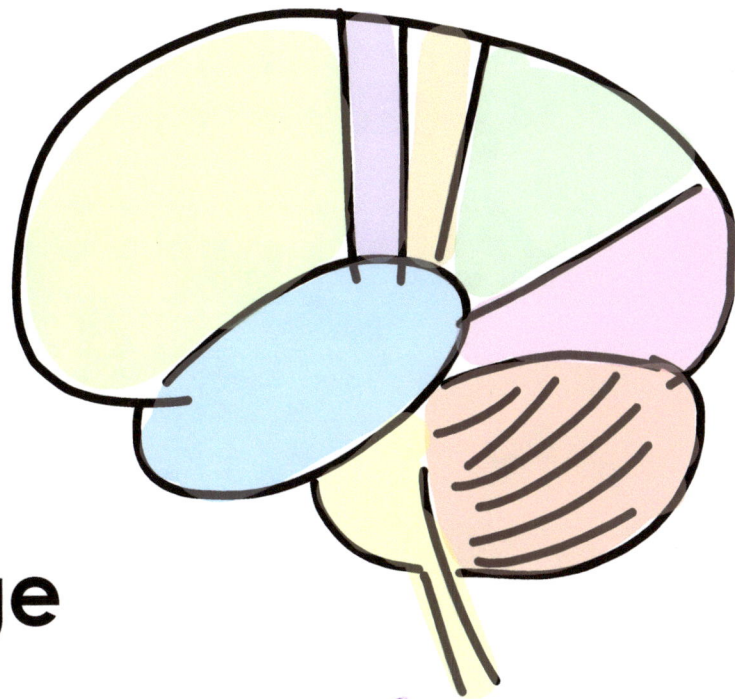

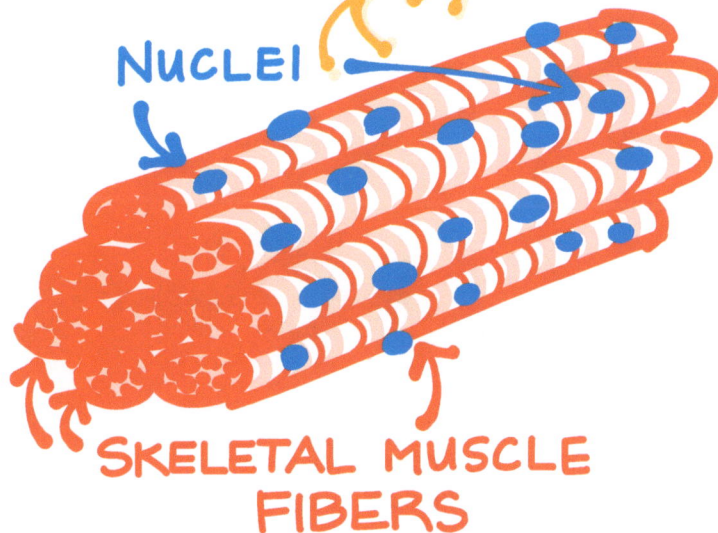

NUCLEI

SKELETAL MUSCLE
FIBERS

Axon Terminal

We are going to look at how the message is sent from the neuron to the muscle.

This happens at the Neuromuscular Junction.

(A junction is a meeting place)

The neuromuscular junction involves this part of the neuron, called the axon terminal, communicating with the muscle fibers.

Some super cool terms to know before we begin:

Calcium (Ca^{2+})
KAL-SEE-UM

Acetylcholine (ACh)
UH-SEE-TiL-KO-LEEN

Receptor
REE-SEP-TER

Synaptic vesicle
SIN-AP-TIK   VES-EH-KL

Sarcolemma
SAR-KO-LEM-UH

Neuromuscular Junction

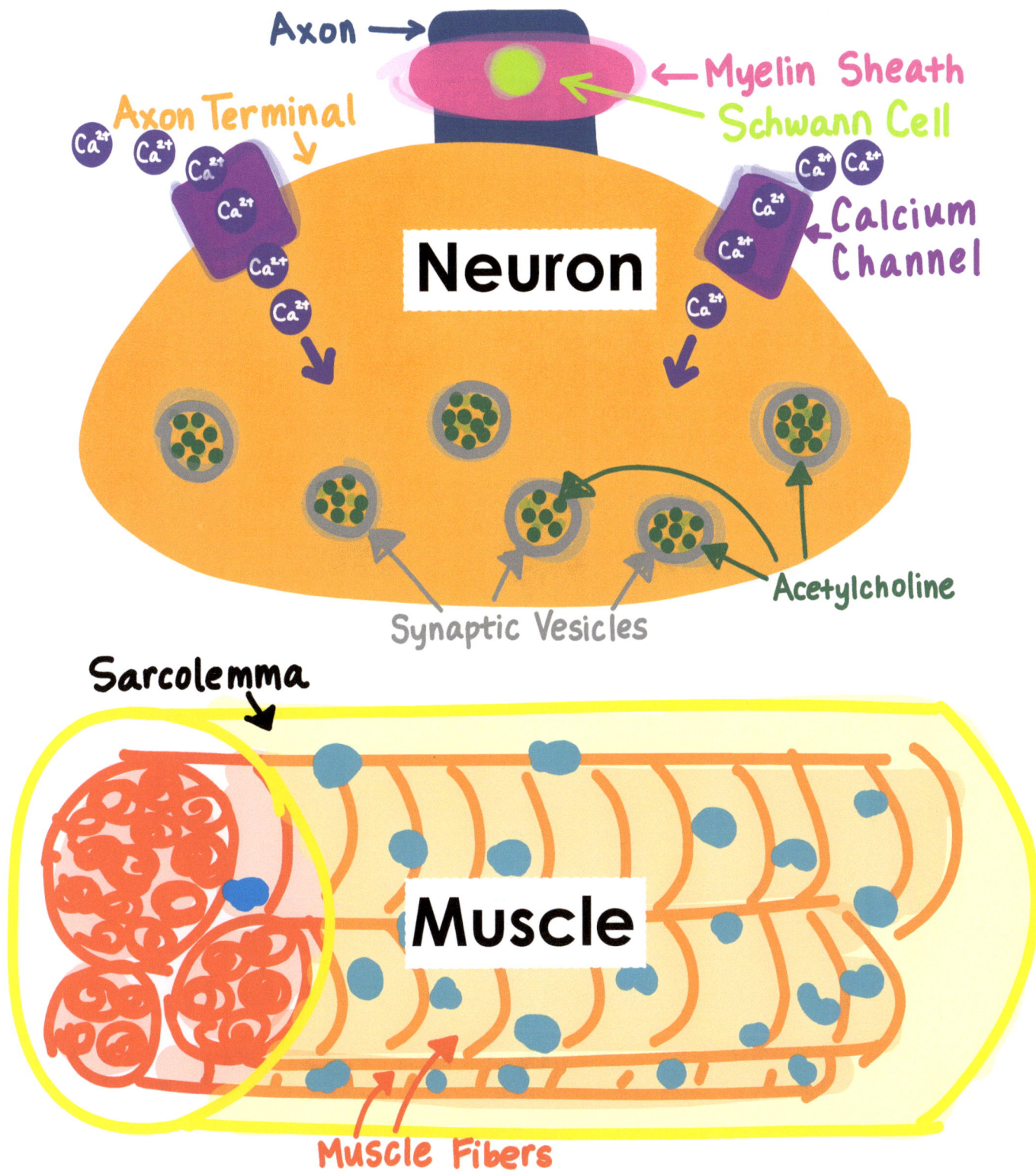

Axon

Myelin Sheath
Schwann Cell

Axon Terminal

Ca²⁺ Ca²⁺ Ca²⁺ Ca²⁺ Ca²⁺ Ca²⁺

Ca²⁺ Ca²⁺

Ca²⁺

Ca²⁺ Ca²⁺

Calcium Channel

Neuron

Synaptic Vesicles

Acetylcholine

Sarcolemma

Muscle

Muscle Fibers

When the message reaches the axon terminal, it allows the element calcium (Ca^{2+}) to enter into the axon terminal.

The entry of calcium (Ca^{2+}) into the axon terminal causes the synaptic vesicles to come out of the axon terminal.

When the synaptic vesicles open, they release acetylcholine (ACh) into the space between the neuron and the sarcolemma (the membrane around the muscle fibers).

Neuromuscular Junction

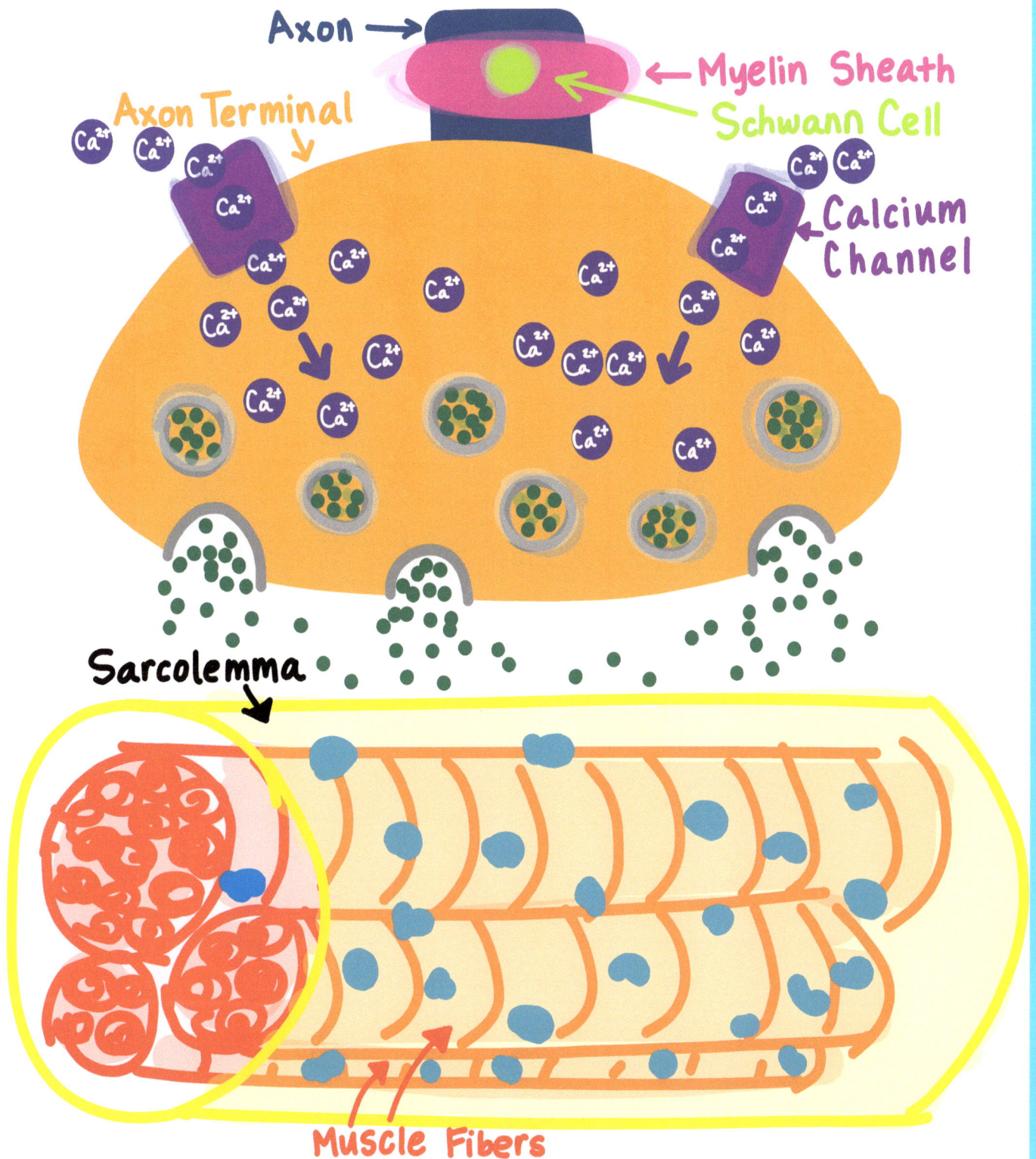

Axon →

← Myelin Sheath

Schwann Cell

Axon Terminal

Ca²⁺ Ca²⁺ Ca²⁺ Ca²⁺

Ca²⁺

Ca²⁺ Ca²⁺

← Calcium Channel

Ca²⁺ Ca²⁺ Ca²⁺ Ca²⁺ Ca²⁺ Ca²⁺ Ca²⁺ Ca²⁺ Ca²⁺ Ca²⁺ Ca²⁺ Ca²⁺ Ca²⁺ Ca²⁺ Ca²⁺ Ca²⁺ Ca²⁺

Sarcolemma

Muscle Fibers

Next, acetylcholine (ACh) binds to the ACh receptors on the sarcolemma.

Review:

The message allows calcium into the axon terminal which causes the synaptic vesicles to release acetylcholine into the space between the neuron and the sarcolemma. Then, ACh binds to the ACh receptors on the sarcolemma.

What is the sarcolemma?
What is the name of this junction where the neuron and muscle fiber meet?

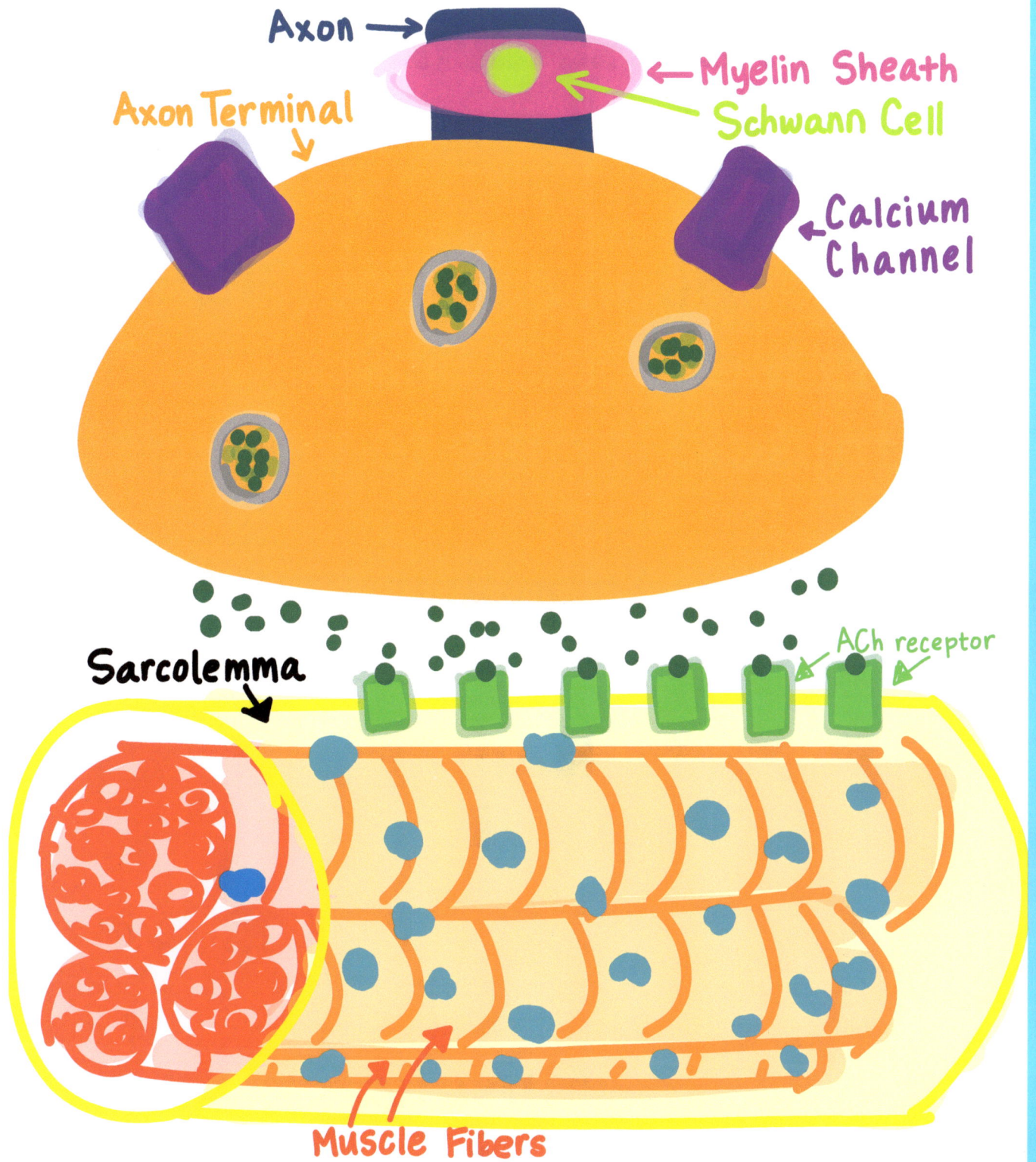

After **ACh** binds to the
ACh receptor, it allows
the element **sodium (Na⁺)**
to enter into the **sarcolemma**.

When enough **sodium (Na⁺)**
enters the **sarcolemma**,
the muscle is able to contract!

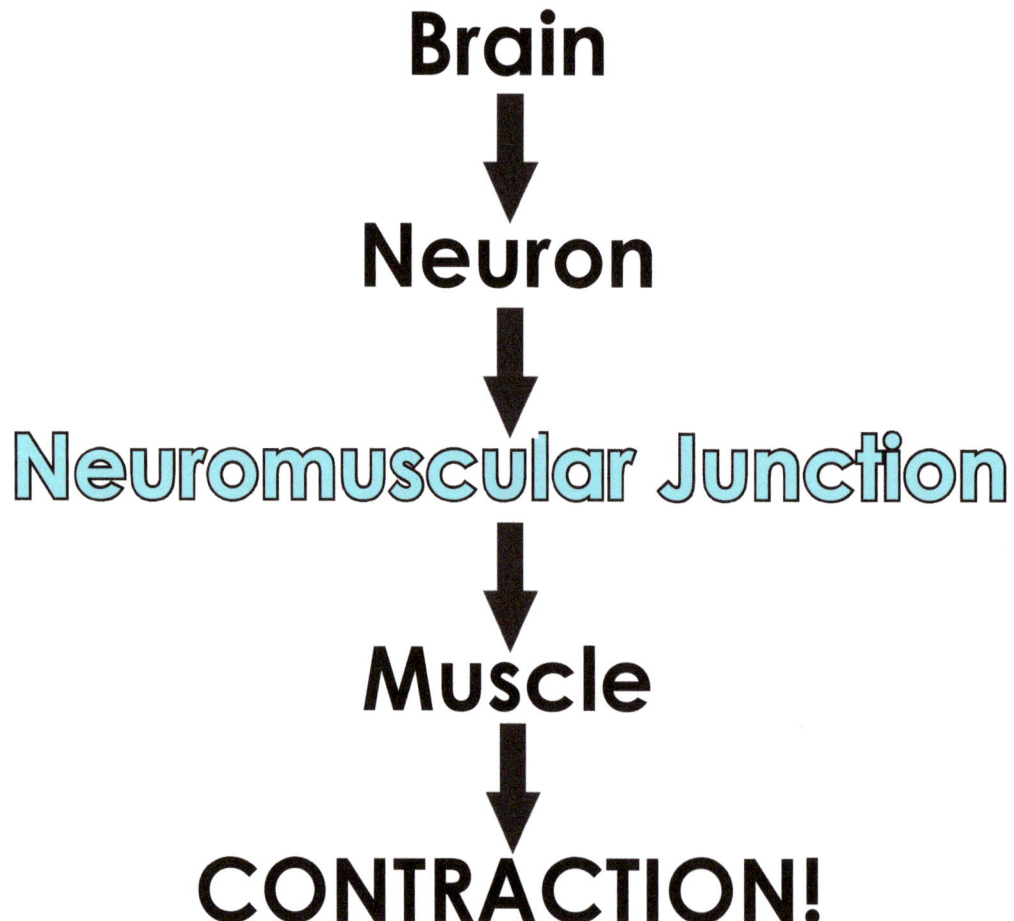

Brain

↓

Neuron

↓

Neuromuscular Junction

↓

Muscle

↓

CONTRACTION!

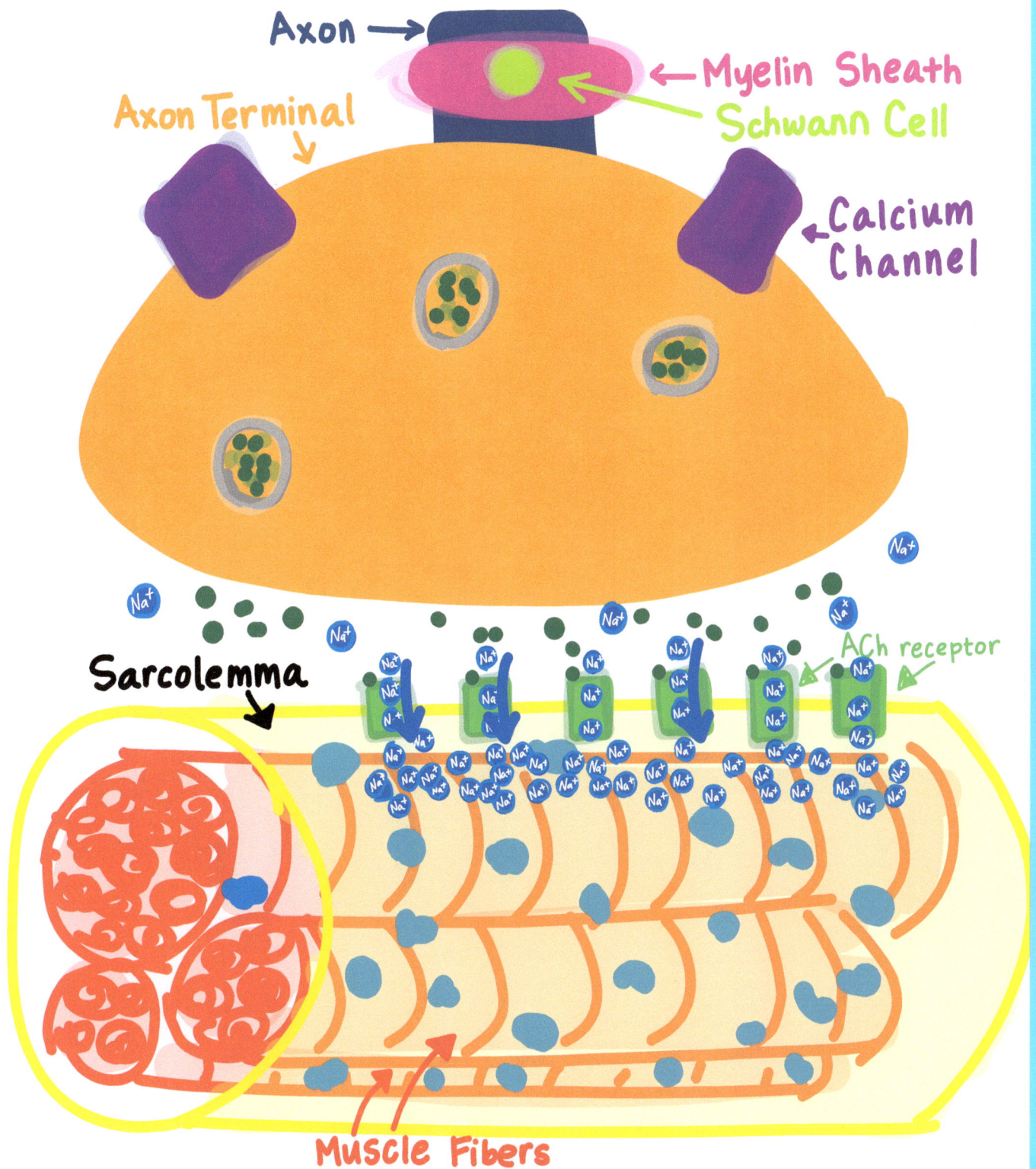

Axon

Axon Terminal

Myelin Sheath
Schwann Cell

Calcium Channel

Na⁺

Na⁺

Na⁺

Na⁺

ACh receptor

Sarcolemma

Muscle Fibers

Sodium enters the sarcolemma, causing the muscle to contract.

A human adult has 206 bones.

The epiphysis is the end of a long bone, the diaphysis is the shaft of a long bone.

The outer compact bone is made of osteons and periosteum.

The inner bone is made of spongy bone and marrow.

Haversian canals go *up and down*. Volkmann canals go *side to side*.

Skeletal muscle is voluntary and striated. Skeletal muscles are attached to bones by tendons.

Cartilage provides structure and cushion in your joints and nose.

Ligaments connect bone to bone.

Cardiac muscle is <u>involuntary</u> and striated from intercalated disks. (ONLY in the heart).

Smooth muscle is <u>involuntary</u> and automatic. (Stomach, eyes).

Movement in our body is started by a message from the brain that travels through neurons to our muscles. The message causes calcium (Ca^{2+}) to enter into the axon terminal, which signals acetylcholine to be released from the synaptic vesicles of the neuron and bind to the ACh receptors on the sarcolemma of the muscle, which allows sodium (Na^+) to enter and cause contraction of the muscle.

New Vocabulary!

Anatomy

Physiology

Bone

Skeleton

Epiphysis

Diaphysis

Periosteum

Compact Bone

Osteon

Marrow

Spongy Bone

Haversian Canal

Volkmann Canal

Skeletal muscle

Cardiac muscle

Smooth muscle

Striated

Tendon

Gastrocnemius

Cartilage

Tibia

Talus

Cuneiforms

Calcaneus

Metatarsals

Phalanges

Ligament

Patella

Voluntary

Involuntary

Intercalated Disk

Neuromuscular Junction

Axon Terminal

Calcium (Ca^{2+})

Acetylcholine (ACh)

Receptor

Synaptic Vesicle

Sarcolemma

Sodium (Na$^+$)

You are a bone, muscle, and neurology expert!

Well done!

You are a Cellular Biology, Chemistry, Neurology, Astronomy & Anatomy EXPERT!